Innovation Capabilities and Economic Development in Open Economies

Smart Innovation Set

coordinated by
Dimitri Uzunidis

Volume 1

Innovation Capabilities and Economic Development in Open Economies

Vanessa Casadella
Zeting Liu
Dimitri Uzunidis

iSTE

WILEY

First published 2015 in Great Britain and the United States by ISTE Ltd and John Wiley & Sons, Inc.

ISTE Ltd
27-37 St George's Road
London SW19 4EU
UK

www.iste.co.uk

John Wiley & Sons, Inc.
111 River Street
Hoboken, NJ 07030
USA

www.wiley.com

Library of Congress Control Number: 2015948060

British Library Cataloguing-in-Publication Data
A CIP record for this book is available from the British Library
ISBN 978-1-84821- 874-1

Contents

Preface

Besides increasing inequality, poverty, hunger, disease and conflict, the impacts on poorer countries of the accelerating pace of technological change, globalization of sociotechnical transitions (energy, ecological and demographic), globalization of technofinancial strategies of firms and the trend toward standardization in the global economic space are problems to which the economic analysis of development must find solutions. Since scientific and technical knowledge of mankind are inexhaustible resources, the question is then how to turn such knowledge into useful goods and services and how to define their usefulness in different socioeconomic backgrounds where these goods and services can be introduced. So, in other words, the question is: what efforts should be made, and for which innovations in developing countries?

Sustained innovation and proactive policy is needed to develop new production techniques and to produce new goods and services to meet local demand, but they must also be sufficiently demanded and reasonably competitive on the global market. The dialectic of development thus supposes that a double performance is needed: innovation to expand the local market; innovation to enter or to stay on the technological trajectories that are needed in the global market.

For now, international technological disparities between countries (industrialized, emerging, middle-income, and less developed) remain significant. What would be the appropriate development models for promoting endogenous technology while simultaneously integrating into the imported knowledge production system? How are technology transfer vectors formed? How do national innovation systems emerge in different contexts and on what institutional foundations? How do policies contribute to organizing production systems? What are the limitations and how do we create the necessary conditions for connecting scientific and technical knowledge with tacit and local knowledge in developing countries? How do we evaluate the performance of such systems?

In this book, we set out to present key intellectual mechanisms – conceptualized by our experience in the field – that support the integration of innovation in general development policy. This book is the result of works carried out within the Research Network on Innovation (http://2ri.eu) whose objective is threefold: to observe and analyze the innovation process, to theorize innovation systems and to promote research in economics and management of innovation. The *Cité des Sciences et de l'Industrie* in Paris, as well as the *Institut CDC pour la Recherche* of the *Caisse des Dépôts* group contributed materially to the production of this book. We would particularly like to thank Lamia Yacoub and Gwenaëlle Otando, the Tunisian and Gabonese researchers, respectively, for their valuable contributions to the study of active economic policies and new institutional architectures in developing countries.

Dimitri UZUNIDIS
President of the Research Network on Innovation
August 2015

Introduction

The study of development refers to standards that characterize innovative capacity, industrial competitiveness and economic performance of rich countries. Development policies therefore aim at bringing countries out of a state of "underdevelopment", which must itself be defined in relation to these standards: widespread poverty, low production potential coupled with high population growth, insufficient promotion of natural resources, survival of an economic and social status described as "archaic", disjointedness of the productive apparatus due to the coexistence of a modern extrovert sector and a traditional survival sector, etc. The question of the constitution of an economic system is paramount: the model, the structure and pace of economic development are schematically determined by political architecture.

With this in mind, economic development is one of the most studied subjects in international economics. The fundamental differences are rooted in the mismatch between two fundamental economic approaches: liberal and interventionist. Development issues fuel a debate of great political acuity that itself refers to an ideological and theoretical debate. Indeed, the requirements and effects of globalization have propelled reflections on the legitimacy of an active economic policy based on innovation to the front of the stage [UZU 10c, UZU 10d, UZU 10e]. The fact is that a proactive role of the State would result in, according to the predominant neoliberal approach, extensive and unjustified interventionism, which is a stark contrast to the context of globalization.

In fact, liberal principles and speeches advocating the neutrality of economic policy were developed around the 1980s with the supremacy of monetarist approaches and then, to some extent, embodied in the Washington Consensus, while Keynesian-inspired regulations were already "well-softened". Recently, liberally-inspired approaches have been developed, which in fact singularly relativize the end of Keynesianism. Today, the idea of the State is still valid, even if its role is challenged in many politico-economic discourses. Indeed, from what the advent of globalization suggests, it is probably unable to consider the organization and development of the world on the basis of a simple interstate system. Hence the recurrent idea of *global economic governance*, which underpins that worldwide regulation is no longer reducible to the exclusive and independent regulation of States, which, as we shall see in this book, reduces the leeway of development stakeholders.

Therefore, at a time when the world economy is mutating and sources of competitiveness are constantly changing, would it be wise to admit that globalization inexorably leads to the substitution of market forces for State interventions in favor of the economy? Otherwise, will a strong foundation for the active role of the State against the hegemony of the liberal approach that underlies globalization be found in recent economic theories (endogenous growth, post-Keynesian growth, neo-institutionalistic growth, etc.)? And so, with reference to theoretical developments as well as recent economic policy practices, what room for maneuver do States actually have to face the challenges of a globalized economy? Is the activation of innovation potential of the

South an economic modernization that would allow these countries to beneficially integrate transnational processes of economic growth?

These legitimate questions, which relate to both theoretical and practical reflections, encourage us to analyze the main elements they raise in the debate on the economic role of active innovation policies in the context of globalization. Our goal is to show that recent development, and seemingly paradoxically, economic policy is nothing other than the implementation of a renewed debate on the proactive role of the State.

The term *globalization* refers both to: the strong integration of national economies into the international flow of capital and goods; and the establishment of international institutions (International Monetary Funds (IMF), World Bank, World Trade Organization (WTO)), a set of rules that ensure freedom of cross-border activities of companies. A transnational legal regulatory framework will tend to fall into place as the result of tensions and political compromises between States. This framework is based on the immutable principle in a market economy where capital must be valued at all times and by all means (financial, commercial and regulatory) with a view to improve potential profit-making. It is now a highlight of globalization that challenges the active and voluntary nature of national economic policies and weakens the role of the State in the development process.

But a certain awareness of the challenges of globalization, and the opportunities and constraints it creates at all levels, seems to be renewing the debate on development issues. Within the context of this debate, there are some for which a global governance system would be the best answer, if not the only answer, to solving the proliferation of interests and conflicting objectives. They rely on arguments drawn from different theoretical approaches (which we will present later) to legitimize this intermediate system between the State and the global market. While for others, the global governance system would be the legitimation of a legal and institutional framework whose true aims are far from achieving the need for new development of the economy and global society. They, in turn, rely on theoretical arguments, but also on historical and recent events, questioning the validity of liberal policies advocated by international institutions. They relate the global governance system to an organizational mode that favors the interests of powerful economies and that embodies the supremacy of decision-making powers of multinational corporations and financial markets [UZU 10b].

While it may be theoretically and logically sustainable, this system seems *a priori* unable to achieve its objective of improving overall performance. Behind this inefficiency, multiple dysfunction problems loom, leading to a crisis of global governance system goals where the global governance system is itself linked to a crisis of legitimacy of recommended actions and stakeholders involved in the decision-making process [UZU 10a, UZU 10b, UZU 10c]. Our goal here is not to propose "ready-made" solutions to this problem, but rather to highlight the gaps in global governance that, until now, have separated us from the perspective of a harmonized, diverse and accelerated development. This analysis will form the first chapter of this book.

The second chapter of this book is dedicated to the study of the main track that developing countries could take to strengthen their economies when confronted with the demands of globalization. The ability to innovate, in fact, allows an economy to improve its performance as well as those of its stakeholders (companies, institutions and individuals). But this capacity depends on the strength of national innovation systems (NISs). The concept of an innovation system is used to measure technological performance and to understand the growth and economic development of a country. The process of innovation and learning from the South differs from those in developed economies. Technology transfer is achieved through the construction of technological capacities and national efforts for absorption and diffusion of foreign knowledge. But this construction also involves the creation of more microeconomic learning processes that are based on indigenous and specific knowledge. The ultimate goal would be to understand the process of innovation in the form of actual construction of skills, with the aim of promoting economic development. But to do this, we will justify the implementation of active innovation policies; we will also discuss the challenges of innovative capacity building in developing countries.

1

Theories and Policies of Economic Development

In the early 1980s, the crisis of state interventionism in developing countries led to the launch of structural adjustment programs (SAPs) recommended by the International Monetary Fund (IMF) and the World Bank. SAPs were considered remedies for the excess debt and the widening macroeconomic imbalances. These programs were implemented by short-term economic policies of monetarist inspiration where the idea of a "minimum State" was the founding perception. SAPs foreshadowed the Washington Consensus, such as was formulated in 1989. It was supposed to be a reference for economic policies that acted most heavily on the structures, advocating reforms of which the nature and meaning seemed consensual. In theory, the objective was to restore macroeconomic stability and especially to promote growth and reduce poverty. But as for the SAPs, a review of the applications of the Consensus does not attest the achievement of its stated objectives. While many economists argue that it was wrong to attach the Consensus to the strict doctrine of liberalism, the fact remains that its implementation was largely inspired from it. The proof: failure of the "minimum State" as a political and ideological objective. By renewing the debate on sources of growth and by "endogenizing" technical progress, new development approaches are advancing theoretical arguments that restore the role of economic policy and, in particular, that of innovation.

Economic, technical and social progress always go hand in hand for many economists. The increase of gross domestic product (GDP) and, consequently, the creation of material wealth translate the improvement of living conditions of the population into quantitative terms. One can compare the GDP of several countries, initially expressed in national currency, in two ways: current exchange rates or purchasing power parity (PPP). To calculate the latter, we use a standard basket of goods then calculate the conversion rate, which is the ratio of the prices of this basket between currencies.

But an increase in GDP does not necessarily indicate improved well–being of the population because its calculation is based on a set of accounting policies. Volunteer work, domestic work or informal economy is not considered in GDP. Hence the concern over the last thirty years is to contruct new indicators that are qualitative rather than quantitative.

The United Nations Development Programme (UNDP) proposed replacing GDP per capita by a composite indicator, the human development index (HDI), which aims to reflect three aspects of economic and social development: (1) life expectancy; (2) level of education; (3) access to the necessary resources to live decently. The level of human development is therefore measured using three indicators: life expectancy at birth (health); knowledge (education); standard of living (adjusted real GDP per capita). The average of the three indices is calculated. The composite index has a value between 0 and 1. A country is classed as developed if its HDI is greater than 0.8; developing countries are, therefore, countries with an HDI of less than 0.8.

The United Nations retain three criteria for defining the least developed countries (LDCs): GDP per capita of less than 900 US dollars; delay in the areas of health, education and nutrition; economic vulnerability: lack of economic diversification, importance of production and export of agricultural products, political instability, etc.

Box 1.1. *Building economic indicators to measure "development"*

Out of all economic take-off models with structural adjustment policies, economic development concerns are perennial [UZU 10a]. The concept of “good governance” renewed the debate on development and the wealth of nations by giving institutions prominent attention. Thus, this chapter’s objective is to show how institutions, and thus the “good governance”, can play an important role in development. As we are aware that we cannot have foresight without a retrospective, we will first look back at theories that marked the development economics up until the 1990s. Subsequently, we will present, by shedding light on works of new institutional economics, the relationship between good governance, global governance and economic development in order to introduce the issue of innovation in proactive economic policies.

1.1. The era of economic interventionism

Development economics dates from the post-war years. On an international level, the decolonization process affected Asia and Africa; the Bretton-Woods institutions were established; the United Nations addressed issues of growth in backward countries, for their industrialization and their need to stabilize prices of raw materials; new regional institutions, such as the Economic Commission for Latin America and the Caribbean (ECLAC), processed regional integration and import substitution strategies. In fact, “developmentalism” (theories about the necessary development of so-called underdeveloped countries), which was formed after the Second World War, mainly drew its references from the economic history of industrialized countries. Therefore, development theories from the time incorporated two major assumptions, besides the paradigm of modernization: the idea that faster growth could only result from the expansion of industrial activities and the idea of voluntarism or intervention in the process of State allocation of resources, “to correct the market laws that had previously distributed industry unevenly throughout the world" [ASS 02, p. 11].

For three decades, development economists correlated the development of Southern countries with state intervention. Theories of economic take-off, which were developed in the 1950–1960s and critical theories of dependence, formalized in the 1970s, focused on state initiative and interstate relations on a global level.

1.1.1. *Impasses of economic take-off theories*

During the first 20 years of its existence (1950–1960), development economics followed the reference model of so-called developed countries to the letter. The dominant post-war Anglo-Saxon economic thinking was in fact Keynesian, or classico-Keynesian. This constituted a reconnection with traditions of classical economic thinking. The dominant economic policies at the time thus gave state activism an undeniable role in the fight against unemployment and achieving growth. At that time, the structuralist approach that developed an analysis in terms of structural parameters (dependence resulting from the primary specialization, etc.) seemed to polarize controversies on development economics, which firstly addressed the problems of underdevelopment to then develop appropriate trajectories.

Underdevelopment was perceived as a series of obstacles for change (lack of capital and entrepreneurs, population pressure, agrarian predominance, weak capacity to innovate). From this point of view, for economic take-off theorists, in order to engage in the path of development, it was appropriate to break the vicious circle of underdevelopment and deploy a sustained and much-needed effort to create enough revenue and thereby increase domestic savings; the latter, by financing new investments, would be able to maintain rapid growth. For A. Lewis [LEW 58], "the central problem of economic theory is to understand the process by which a community that was previously saving and investing 4–5% or less of its income, turns into an economy where voluntary saving is about 12–15% or more of its income". Indeed, the first theories of development, formulated in the 1950s, advocated the image of the Marshall plan: large international financial transfers to third world countries, to enable them to accumulate the necessary capital for a critical investment threshold in order to initiate accelerated industrial modernization [SAW 87].

According to Rostow [ROS 60], take-off is a transitional phase of about twenty years after which growth moves toward maturity and then to mass consumption and finally, to a more moderate growth. This theory was criticized for several reasons: vagueness of the periodization, imprecision about details for setting up favorable conditions for take-off, excessive trust in the power of a "centralized and efficient" State [BIE 06]. Moreover, colonization was positively perceived as having laid the foundation for turning a traditional society to a modern society.

Ultimately, in the tradition of the classical economists (Smith, Ricardo and Malthus), the development economics of the 1950s focused on accumulation and reproduction. Insofar that capital was the factor preventing economic development, priority was given to the savings rate, the investment rate and the choice of techniques in line with the availability of two main production factors: labor and capital. Development thus became evidence of finance. From this financial injection, economists expected to break the vicious circle of poverty ([NUR 53] according to which poverty fuels poverty), to accelerate the massive transfer of labor from agriculture to industry (Lewis theory), to initiate industrial growth [ROS 43] and, more generally, to trigger the transition of the society toward the industrial era.

The path that was taken allows us to distinguish between liberal Anglo-Saxon economists who assimilated the role of the State to optimal allocation of resources in an open economy, and European heterodox economists, who recommended increased state intervention in a protected industry. Indeed, the central idea of the latter was that third world States should opt for *selective voluntary public investment* [PET 98, pp. 14–20] in favor of industrial sectors that were considered strategic in terms of economic benefits. This gave rise to a range of theories: (1) ripple effects [HIR 74], (2) growth poles [PER 59], (3) industrializing industries [DES 71]. However, despite the enormous differences between these theories, they focused on one main idea: "The developmentalist voluntarism of the state elites in the third world then seemed both obvious and unequivocal to solve the issue of socioeconomic development without any political problem" [PET 98, p. 15]. Indeed, these theorists were overly trusting in the State and showed no skepticism toward it. Therefore, the question was not to know whether the State was effective or not.

That said, in the community of development economists, only the Swede Myrdal [MYR 57] raised the question of the nature of the State in poor countries as a possible obstacle for development. He insisted, therefore, on the risk of the existence of a State either too soft to conduct efficient policies, or too authoritarian or too corrupt to carry out appropriate policies for redistributing the fruits of growth. However, this warning against the deviant political behavior of the leaders was overshadowed by mainstream economists who threw these political issues out of the economic field.

In the end, the development economics in the 1950s was based on the principle of rationality of the State in the long-term. Indeed, the rationalist ideal that has dominated Western thinking since the 19th century greatly influenced the first nationalist elite of the Third World borne from the independence movements. The "dependency theory" was then a reaction against the "structuralist" reformist current and against the evolutionary patterns, which it generally equated with liberal thinking. It mainly criticized the "*compradores*" elites, who equated their interests with those of the elites of industrialized countries. It also denounced the dominant discourse on state voluntarism as "neutral". Such thinking, influenced by the Latin American current and especially by the thinking of Prebisch [PRE 49], saw capitalism as the determining factor of underdevelopment and usually rejected the modernization agenda for a break with the international market in favor of import substitution by local production.

Underdevelopment was no longer defined as a developmental delay but rather as a product of the dynamics of capitalism on a global scale [AMI 73]. There was unequal exchange between developed and underdeveloped countries [EMM 69]. The development economics was a prisoner from birth of the context of international relations at the time. Thus, the specialization of such countries in the production of poorly developed raw materials quickly attracted the attention of UN experts. It gave rise to the publication of a report written by R. Prebisch in 1949 [PRE 49]. This document discussed the difficulties faced by third world countries in transforming and adding value to their natural resources locally. Moreover, as the prices of the natural resources tended to decline compared to manufactured goods, the report radically challenged the specialization model based on the comparative advantages.

The inclusion of structural aspects in the analysis of Southern economies was of utmost importance. Underdevelopment was not analyzed as a natural phenomenon, but rather as a historical situation related to the disintegration of productive structures and dependency phenomena maintained in the international economy.

Indeed, the gap between elasticities and the limited number of products exported by the periphery was the origin of the secular deterioration of terms of trade. In this context, "dependency" economists made a series of recommendations to break the vicious circle of underdevelopment. Thus emerged strategies proposed to replace imports with local production. The focus of development on the domestic market and state intervention should have allowed a reversal of the trend toward unequal development between the center and the periphery. However, the stagnation of Latin American economies, and all countries that adopted the import substitution strategy, were the source of early criticism. Some authors, like Furtado, blamed the failure or perversion of these policies on the unfavorable integration of these countries into international trade: "More subtle and insidious forms of dependence, infiltrated in our financial and technological circuits, came to replace the supervision previously practiced by external markets on the regulation of our productive activities" [FUR 95, p. 63].

Despite the relevance of their analysis, which gave the study of structures a prominent role, these theories either underestimated or overestimated the role of the State. On the one hand, indeed, the place given to the State for correcting market imperfections and for designing public policies seemed significant. But the question of the behavior of the elite was hidden or overshadowed. For example, analyses by S. Amin [AMI 73] completely ignored the role of politics. The latter was reduced to an instrument in the hands of foreign interests. And so, the State, in this framework, was a puppet structure. Moreover, Furtado, by allowing the State to play an important role remained aware of the risk of perversion of development strategies in the context of the proliferation of coups d'état. As such, the advent of a new ruling class driven by the search for individual interests began. This is in line with the recent analyses by Stiglitz, for whom the distribution of wealth in some developing countries is not determined by careful trade-offs between equality and efficiency: "It is not defined under the principles of social justice; it is the result of brute force. Wealth is power, and this power allows the ruling class to keep the wealth" [STI 06, p. 198]. Thus, we find ourselves at the heart of the institutional performance paradigm.

1.1.2. *The crisis of the interventionist State*

The 1929 crisis gave rise to a particular echo in analyses by John Maynard Keynes (1883–1946). The leading thread of his ideas was that for market economies experiencing sustainable endogenous imbalances, the onus shifts to the State to support growth and stimulate the economy to achieve full employment. Its intervention should be done via economic policies for boosting demand and by committing to additional public expenditure, corresponding to the regulatory function of the State. Concomitantly, to avoid a return to protectionism, the Bretton-Woods agreements adopted the principle of progression toward international liberalization through cooperation between external policies. And to avoid returning to unemployment and the inequalities of the 1930s, they left a margin of freedom for internal policies of full employment and the welfare State [COU 03].

Thus, after the Second World War, the States strengthened their role in the economy. The priority was to correct market failures, especially macroeconomic ones: boost growth, promote full employment and external balance, ensure price stability, improve living standards, etc. At the time, international exchange began to intensify, but growth depended mainly on the dynamics of domestic markets. This heightened the importance of the nation-State through a high degree of political, social and economic interactions that occurred internally. The trend was therefore toward national integration: consolidation of the welfare State in the first half of the 20th Century strengthened the legitimacy of the central government and, consequently, justified its massive intervention in the economy.

The global economy changed dramatically in the 1970s, in response to the changing practices of policy makers, who, under the pressure of circumstances and public opinion, increased interventions in the economy: external deficits and international debt, stagnation and inflation (*stagflation*), corporate bankruptcies, spiraling unemployment, etc. More generally, an area of economic "turbulence" occurred, but which supporters of the liberal theory were slow to name "crisis" [HUM 95]. Since then, Keynesian-inspired regulations have been "dampened" after more than thirty years of economic and interventionist practice.

It was the welfare State system that created a context of international crisis, conducive to neoliberal discourse, in the early 1980s. This was triggered by a slowdown in growth, growing imbalances, practice of social assistance and especially "hysterical" public spending. The arguments of liberal economists undermined the welfare State, first in England, then in the United States and then finally, triggering a liberal wave that gradually spread to many countries that, for the most part, also became supporters of "less State intervention". The problem was simple: as a good policeman, the State must intervene to create and enforce conditions for effective competition through market transparency. Public or private monopoly, substitution of the entrepreneur by the State, laws and social protection of employees, etc. were obstacles to innovation, to the detriment of consumers and employment. The role of the State was thus to ensure fair competition, to avoid excessive concentrations of economic activities and to protect consumers.

The turning point in the functioning of national economies and the global economy occurred in the 1980s. In 1979, the arrival in power of Margaret Thatcher in Great Britain and, in 1981, of Ronald Reagan in the United States, facilitated the advent and spread of liberal doctrines; in his inaugural speech, Ronald Reagan said that "the State is not the solution, it is the problem". In the late 1970s, Senegal inaugurated the first SAP, the debt crisis having begun in developing countries thus forcing them to adopt "market-friendly" development strategies. This unification of economic models also won over the Eastern countries: in 1984, China opened its first special economic zones. Five years later, the fall of the Berlin Wall announced the liberalization of the Soviet Union in 1991, which was also the year when India, who had been nationalist until then, in turn liberalized itself. Thus, in ten years, the configuration of the world changed dramatically with the rise of liberal-inspired approaches; approaches that were largely concretized through widespread *therapy* of SAPs for countries in crisis.

1.2. The era of liberalism

Development economics were completely transformed from the 1980s. Debt crises were reorienting priorities. Thus emerged the need for balance, which expelled the temporal dimension of change. Due to this, the plurality of theories shrunk in favor of the liberal theory to which some neo-Keynesians aspects were added. However, the nagging question remained: why did some developing countries achieve good results in terms of development in the post-colonial period and others stagnated or even regressed? Differences in economic policies played an important role. In fact, international institutions strived to recommend a combination of economic policies grouped under the name of SAPs to developing countries; these advocated the establishment of a sound macroeconomic policy, the liberalization of domestic markets, reduced State spending, integration into the global economy, etc. The recommendations in these programs were invariable; specificities of developing countries, which fueled the early work on development economics and which were initiated by structuralist economists, were left out of the discourse of international institutions. The simple operation of market forces in a context of free competition and free insertion of the national economy into the world economy, was thus supposed to guarantee that poor countries caught up with the richest countries in terms of economic prosperity and social well-being.

However, structural adjustment was not intended to assist in the creation of internal economic dynamics. That was for the authorities of countries that were subjected to the adjustment to deal with. The role of the adjustment was to ensure that the evolution of the balance of payments allocated the necessary resources to paying of a debt. The economic situation that was consistent with this observation was that activities involved in the domestic market slowed down, wages were compromised, indirect taxes were high while direct taxes were low and currencies were subjected to competitive devaluations. In this new vision of economic development, the State had to seek macroeconomic stabilization and refrain from interfering negatively with market rules.

1.2.1. *Structural adjustment programs*

Structural adjustment programs were set up to overcome the interventionist State crisis, which was manifested by a marked deterioration of internal and external balances, a result of unsustainable protectionist, inflationist and fiscal policies, particularly in Latin America, but also in varying forms in Africa and Asia. The second oil crisis worsened the situation, which struck developing countries more heavily than developed countries, which benefited from the recycling of petrodollars. The higher interest rates on international financial markets and deteriorating terms of trade resulted in widespread and widening imbalances. In the early 1980s, most developing countries faced the problem of international insolvency and cessation of payments.

Sitting on their role as funders, the Bretton-Woods institutions advocated national economic policy guidelines for liberalizing their economies toward hypothetical deleveraging. SAPs were presented as a therapy of liberal inspiration, strengthening market mechanisms, improving productive and commercial efficiency and reducing the discretionary power of the State.

To begin with, the priority was given to the consolidation of public finances and the easing of bureaucratic apparatuses, at a time when the level of demand was considered to be too high, the investment and consumption were based too heavily on imports, the inflation was too high, the weight of debt was too heavy, the economic competitiveness was too weak and the economies were not export-oriented enough. Interventionist economic policies were considered responsible for causing distortions. It was then necessary to enhance economic efficiency and ensure non-inflationary growth through the withdrawal of the State in favor of the free operation of market mechanisms.

In the logic of the IMF and the World Bank, the SAPs should have been twofold: stabilization, then the adjustment of structures. The IMF's actions corresponded to the first part, being more focused on monetary and financial aspects and presented as short-term stabilization policies of demand and recovery of a viable balance of payments. The actions of the World Bank rather corresponded to the second part, being more oriented toward the structural policy of supply management and modification of the conditions of production.

The "stabilization" was based on measures that were as diverse as reducing public spending, slowing the increase in money supply by limiting credits to the economy, raising the interest rates, subordinating wage developments to productivity growth, devaluating the currency, etc. It was the typical policy mixes recommended by the international financial institutions, particularly the IMF before theorizing and advocating the practice of monetarism in an open economy. Let us recall that the monetarist current was born in the late 1940s, driven by M. Friedman [FRI 53, FRI 69] as a reaction against Keynesian preference for fiscal policy. Monetarist theories proposed a restriction of State intervention and believed that inflation was due to an excessive volume of money circulating in the economy; hence the need to implement a monetary policy that restricted the money supply.

The "structural adjustment" aimed to create the conditions for stable and sustainable economic recovery to ensure balanced growth via liberalization of productive, financial and business systems: gradual liberalization of imports, prices and interest rates, easing of public control over private national and foreign investment, restriction of direct government support to companies through the liberal percept *trade not aid*, withdrawal of the State from the productive fabric by massive privatization of public corporations, etc.

Indeed, perceived by the liberal approach and the agency theory as a means for achieving specific interests, the public company was subjected to strong criticism, on behalf of the superior efficiency of coordination mechanisms by the market: public production did not maximize public interest, but rather private interests, at the expense of economic and social waste. This notion of "waste" highlighted the distortions that State intervention could create and it was then recommended that economic policy be neutral and that liberalism be the rule. In this sense, *laisser-faire*, market regulatory mechanisms and the benevolence of the invisible hand were to ensure optimal operation of the economic system.

It was according to this same logic of *minimum State* that fiscal and financial reforms increased and the privatization process was accelerated over the past three decades in developing countries but also in the United States, Europe and in former socialist countries. The results have not matched expectations: in the late 1980s, the mandated treatment plunged developing countries into recession. Long-term development needs were side-lined. A number of developing country governments continued to implement the SAP, most for fear of suspension of loans and financial isolation, rather than for conviction of their merits or their success. At the end of the 1980s, the World Bank [WOR 87, p. 2] itself recognized that for many developing countries, "the adjustment proved to be a more lengthy process than was envisioned". The SAPs were thus criticized for having too strict a vision of conditionality, but in reality, and at least until the early 1990s, there were a lot of loans and very few adjustments.

Despite the mixed results, for international financial institutions the neutral economic policy and the logic of a lesser State were still desirable in the long-term; the problem lies deep in the configuration, the rhythm and the steps toward liberalization. It was apparently around this logic that the ten commandments of the "Washington Consensus" were founded, as formalized in 1989 (see Box 1.2).

– "*Fiscal Discipline*": whereas public deficit is a source of inflation and external deficit, fiscal austerity aims to deleverage the State, but also maintain and improve purchasing power, mainly in the categories of the most disadvantaged populations.

– "*Redefining priorities in public expenditure*": subsidies to the economy, for employment and to enterprises must replace direct aid funding health, education and infrastructure construction.

– "*Fiscal reform*": to counter tax evasion and the rise of an informal economy, but also to improve the finances of the State and give new life to the economy, the government must pursue two objectives: expand the tax base and lower marginal tax rates.

– "*Liberalization of interest rates*": the market must set interest rates, but the State must ensure that these are positive and moderate in order to be attractive to international investors. These can contribute to financing development.

– "*Competitive exchange rates*": the aim is to promote exports. Controlled currency depreciation should move in this direction while avoiding inflationary spiraling due to excessively low levels.

– "*Trade liberalization*": export promotion cannot be achieved without the liberalization of trade; to limit or even remove tariff and non-tariff barriers.

– "*Liberalization of direct investment from the outside*": first, foreign investment must be unfettered, then the international financial institutions imposed the liberalization of the movement of all kinds of capital, leading to significant financial crises throughout the 1990s.

– "*Privatization*": reduce public deficit, contain state intervention, but also make it more competitive (in liberalized markets) through more appropriate management, these are the main objectives of privatization that receive the broad consensus of experts of the "Washington Consensus".

– "*Deregulation*": contestability of markets must be applied on a large scale. The supposed success of this policy in the United States (Reagan years) should inspire all governments; to eliminate barriers for entry and exit of the markets and promote free enterprise.

– "*Ownership rights*": the reinforcement of property rights fosters individual initiative and allows the informal sector to obtain ownership titles at acceptable costs.

Box 1.2. *The "Washington Consensus" in ten points*

The Washington Consensus derives its name from an article by John Williamson in 1989. It is in Washington that the headquarters of the IMF and the World Bank, the US Treasury Department and many influential think tanks like the Institute for International Economics are located. In this expression, Williamson defines "Washington" as Washington politics of Congress, as senior officials of the administration and technocratic Washington of the international financial institutions, of governmental economic agencies, of the Federal Reserve Board and of think tanks [WIL 90]. His philosophy remained the same as that of the SAPs: strict budgetary discipline, strict monetary policy, openness to international trade. The idea was that the revival of economies, often ossified by bloated administrations, should be carried out by a supply policy: encouraging private investment and consumption with tax concessions and lower rates of direct taxation.

1.2.2. *Failure of the "minimum State"*

In the early 1990s, Latin American States, like many other developing countries, made the Washington Consensus their cause. Clift [CLI 03] pointed out that this held some of its promises: stronger budgets, lower inflation and debt ratios, an influx of foreign investment and a recovery in growth. At the same time, unemployment rose and poverty remained endemic, while the opening markets exposed these countries to the collateral effects of globalization, including the influx of speculative financial capital, which was higher than foreign direct investment (FDI).

On a theoretical level, but also in practice, the Washington Consensus was quickly challenged. According to Stiglitz [STI 98], the framework offered too few instruments, a restrictive vision of development and senseless marginalization of the role of the State. Moreover, it has often been held that the withdrawal of the State from its role as producer via massive privatizations, combined with deregulation and liberalization of FDI (if these reduce the budget deficit and boost growth), induces the strengthening of economic dependence of countries whose production bases and innovation systems are weak [STI 02, BER 04, UZU 05]. Moreover, Rodrik [ROD 98] showed that the liberalization of capital flows does not lead to more sustained growth and development because of their volatile financial capital that sought immediate payment.

Financial liberalization was supposed to attract international investment capital in areas where the comparative advantage of a particular Southern country was identified and thus enabled the country to benefit from non-debt-generating funding. But not only this, because the opening and strengthening of attractiveness vis-à-vis the FDI were a factor of technology transfer from the North to the South and a means for reducing unemployment. However, capital flows proved to be highly concentrated and volatile, mostly corresponding to speculative investments (not to actual investment). This was where *financialization* of the global economy resided and created a kind of dichotomy between the real and financial spheres. And we do not take unnecessary risks on considering that international institutions contributed by advocating and imposing liberal policies based on the primacy of the opening of the capital account. Many economists now belatedly note that the sequencing of reforms matters enormously and that countries should first strengthen state control. Paradoxically, the institutions of this global governance [UZU 10c] had the greatest difficulties facing these markets as they were unable to predict or explain the upheavals and movements that were contrary to the classical theory of liberalism that had inspired them so.

Indeed, despite progress over more than half a century in the understanding of economic processes, and despite efforts from the IMF and other international stakeholders for effective governance of globalization, worldwide crises have been more recurrent and, in some ways, more serious. The financial crises that have erupted over the past decade are representative examples, such as those that occurred in Asia (1997), Latin America (Mexico, 1995; Argentina, 2002), in Russia and in 2008 in the United States followed by the rest of the world. Violent and costly crises aggravated poverty and inequality, relatively and to varying degrees, to the point that we can today speak of a "globalization" of poverty. Each time, these were opportunities to renew questions and revive reflections on the role, architecture and efficiency of the global accumulation framework, which was neither capable of stabilizing the global economy, nor effectively preventing and mitigating crises. Stiglitz's [STI 02] reflections on this topic always proved to be valid: when a country is in crisis, not only have the IMF funds and prescriptions failed to stabilize the situation, but in many cases, they made them worse. It is indisputable that the IMF failed in its original mission to promote global stability and was not more "brilliant" in the new tasks it set, for example to guide the transition of ex-socialist countries toward a market economy.

Financial globalization required mature financial systems, but not only this, as state regulation was also imperative for measuring the risks of capital mobility. The failures of globalization were due to the supremacy given to the private sector in the recommended model of development, gradually confining the crucially important role of the State to mere regulatory functions. It was not surprising then, that the influence of the private sector extended increasingly, and by default, in areas where State authority was weakened.

Indeed, developing countries were led to practice pro-cyclical policies as prescribed in the Washington Consensus but which, in fact, worsened their situation: "Countries of the developing world have been asking why the United

States, when faced with an economic crisis, is in favor of expansionary fiscal and monetary policies, whereas when these countries themselves are in the same situation, they are required to do exactly the opposite" [STI 02, p. 308]. Developing countries were not in a position to counter the commandments of consensus. The restructuring undertaking became so complex that they often seemed politically and socially untenable. The need to eliminate all barriers to trade, investment and currency transactions was sharply opposed to the idea that these countries had to protect, consolidate or strengthen their economies. In the late 1990s, the results turned out to be different from what Williamson had expected, that the IMF and the World Bank had promised, what the States had hoped for and that econometric models had predicted. Was it because the liberal precepts on which the Consensus was based were poorly adapted to the real requirements of the context and national conditions? Or was it because there was a misinterpretation of the ten commandments?

In the second half of the 1990s, ten middle-income countries experienced serious financial crises with consequences such as the decline of exports, accelerated deindustrialization, massive loss of employment, etc. The Asian crisis was a clear example, having occurred due to the excessive trust in the market and commitment toward deregulation based on the "minimum State". On the other hand, China and India, who experienced sustained economic growth, only gradually opened up and maintained a proactive economic policy. Chile having led an ultra-liberal economic policy in the 1970s, its performances thereafter were the result of a turn around to more economic voluntarism from the mid-1980s. These countries adopted counter-cyclical policies: support for exports, strengthening the system of education and research and control of short-term capital inflows.

1.3. The era of "good governance"

After the Asian crisis in the late 1990s, it appeared that the consensus raised more problems than it solved. A "post-consensus" seemed to be emerging, in which the idea of a minimum State was put into perspective, at least regarding the pace of its withdrawal, the distribution of the fruits of growth and, ostensibly, the central role of institutions. In its 1997 annual report, the World Bank inflected its position by stating that a "good governance" was imperative for the proper functioning of the market. Rogoff [ROG 02] considered that the negative effects of the consensus were due to "bad governance" and poor conduct of economic policies. Rodrik [ROD 03] proposed to expand the Washington Consensus, focusing in this case on good governance.

The failure of the structural adjustment policies was interpreted by the World Bank as a lack of institutional capacity in some countries, so it directed its programs to what it called "the good governance". Therefore, the policy made a remarkable entry into the discourse of international institutions. In fact, in addition to the emphasis on the implementation of the programs, it was also about understanding the nature of the institutions that embodied these programs. Thus "good governance" could, *a priori*, refer to the inclusion of political behavior in development. That is to say, political cost was highlighted in the new theories of development.

According to the new discourses, those States with a rational institutional architecture were the most capable to promote development. Institutionalism thus burst into the debate on development. From then on, the focus was on the institutional deficit that afflicted developing countries. In other words, if some theoretical receipts did not prove to be effective at the empirical level, it was because the institutional architecture of the economies in which they were applied was lacking. In this section, we propose to trace the outlines of a new development model by placing the question of institutions at its center. The *a priori* that governs this analysis emphasizes that the existence of good quality institutions is both the result and the cause of economic prosperity. However, regardless of causality, increasingly abundant empirical research shows that institutions have an important effect on determining the allocation of production resources and income distribution [ROD 05].

1.3.1. *Institutions, "good governance" and development*

In the 1990s, the dominant paradigm of development economics changed. The symbolic failures of all States (planning) and all markets (minimal State) led to a metamorphosis of development economics. The theory of "good governance" took over. This theory assumed that there was a strong complementarity between democracy and the market, in the sense that political and economic systems were mutually reinforcing each other [FIT 04]. The issue of institutions became crucial. The report of the World Bank in 1991 was indicative of these changes: the poor performance of certain countries was explained by the quality of "institutions". The institutional deficit experienced by the economies of developing countries was put forward to explain the gap in the economic performance with the North. In essence, development was not only conditioned by factor endowments. The institutional component explained a part.

The concept of "good governance" is now ubiquitous in economic analysis. Proponents of this concept present it as a healthy alternative to power abuse in its current manifestations, a cure for all ills of contemporary society and especially the optimal way to ensure development in countries suffering from endemic underdevelopment. Its adoption by international institutions has allowed us to highlight some changes in the status of the State in economic theory of development. *Bad development* results from *bad governance*.

Williamson [WIL 94] was considered the father of *new institutional economics* based on the assumption of bounded rationality and opportunism of agents. He distinguished between different types of institutions, such as the market, the hierarchy and the hybrid forms. The central idea of the new institutional economics was that institutions mattered significantly in economic processes and could be analyzed using neoclassical theory tools. The inclusion of institutions thus represented a major advance in liberal theory.

D. North [NOR 90] perceived institutions as the rules. He considered that there were two kinds of institutions: formal rules (constitutional rules, rules of property rights and contracts) and informal rules (standards and practices). The definition of institutions developed by North was meaningful insofar as institutions were seen as the rules of play of a society or, more formally, the constraints defined by men to shape their interactions. In this way, institutions ensured that the rules were respected in a context where different types of transactions occurred repeatedly. They had characteristics of public goods that the market could not supply efficiently. It is in the logic of minimizing transaction costs, in the control of opportunistic behavior within a relationship or in the reiterated balance that the explanation for the emergence and functionality of institutions can be found. Institutions are then based on power.

Furthermore, if the idea of a combination of institutional architecture and economic performance seemed common to several economists, North insisted that it was incentives that served as a mediator between institutions and economic performance. Institutional framework determines the behavior of stakeholders; similarly, stakeholders will be the source of institutional change. The organizations and stakeholders that emerged seized the opportunities created by the institutional framework. If this rewarded speculation, speculative organizations would appear; if it rewarded innovation, innovative businesses would be created and systemic innovation would begin.

For the World Bank, governance is "the set of rules governing the exercise of authority in the name of an electorate comprising of selecting and replacing those who exercise this authority" [WOR 09] and good governance is to exercise this authority by respecting the integrity, rights and needs of everyone in the State. Also according to the World Bank, relations of good governance in a framework based on two universal values can be considered: social inclusion and responsibility. The idea is that insufficient quality of governance blights the economic, social and human development; which explains the economic delay in Southern countries compared with Northern countries.

Good governance applied to developing countries required, according to the World Bank, the development of education and infrastructure, environmental protection and equitable distribution of resources as the necessary conditions for markets to function properly. A system of laws was necessary to regulate the liberalization of the markets of product, capital and labor in order to avoid the excess of capital flight and the increase of illegal and informal activities. Then, institutional reform was needed to better monitor the economy and enlist all economic stakeholders (political, business and trade unions) in the decision-making process. Finally, the tax system should ensure the proper distribution of income. But it also had to ensure that the poor "have access to assets": instruction, ownership titles, microcredit, land reform, etc. It was not a question of returning to the hypertrophied, corrupt and expensive State, but rather moving toward an "astute State". Institutional reforms aimed to implement a good decision-making process through the adoption of good policies.

Let us note however that the concept of governance is generally (on a theoretical level) and particularly (by examining the facts) questionable. It is imbued with a strong dose of authoritarianism [UZU 10c, UZU 10d]. Indeed, if one refers to the question of institutions as a common infrastructure ("common good") to all agents of a national economy, governance and democracy do not necessarily go together. In this case, "democracy" (deliberative) is not linked to any "common destiny" or legitimacy given to citizens' acts as individuals, but it results from the intervention of "stakeholders": dialogue of governments with economically or financially powerful forces (firms, banks, etc.) or with influence (lobby groups, unions, religions, etc.). The "moral" perspective of a "common good" is then substituted by a "political" perspective of the definition of "good". These institutions that are formed by power relationships (of conflict and cooperation) create an economic development trajectory on which all stakeholders of the economy operate (or must operate to avoid exclusion).

Ultimately, from the World Bank's approach, liberal thought tried to promote a model of organization of developing economies based on the idea that democracy and the market are not mutually exclusive. Therefore, the two complement each other. From this perspective, the State was expected to play an increasingly significant role in development through the establishment of an infrastructure and institutional basis. This explains the importance that researchers of development economics and stakeholders of the international community give to the topic of institutions. According to these stakeholders, it is the principle of effectiveness that guides the selection of institutions. But this principle of effectiveness is itself defined by the power relations that create the overall framework for the governance of the global economy.

1.3.2. *"Development" in global governance*

The growing assertion of failures of liberalism (concretized by rather mixed results of the SAPs and the applications of the Washington Consensus), combined with the needs of new economic theories of innovation that showed the important role of institutions, organizations and their interactions, shifted the development debate toward the conduct of "clever" policies to which national and international stakeholders had to adhere (grouped in a donor system: banks, companies, NGOs, IMF, World Bank, foreign governments, etc.). Developing countries then found themselves caught in a dialectical relationship between their own "governance" and governance of the global economy.

Indeed, the rise of the topic of globalization and the consequent challenges for the nation-State suggested, according to functionalist logic, a transfer of regulatory instruments that had lost their effectiveness on a national level to a global scale. In other words, it was a new model of representation and management of the interdependence that should have emerged and been applied to a growing number of areas, in response to growing constraints and global problems arising from globalization. This globalization, often perceived as a process of homogenization of public management, rather proved to act as an accentuation factor of differences of all types and at all scales.

A system of global governance should thus be the real place of power, faced with increasing complaints against reforms from developing countries and demands of civil society, which then join together to challenge the influence of developed countries and large firms and institutions. The aspiration to a more balanced and equitable globalization seemed to begin to materialize in the creation, at the end of the Uruguay Round, of the World Trade Organization (WTO) and its Dispute Settlement Body (DSB). This symbolized the assertion of an arbitral power that relativized that of the most powerful States whose practice would be governed by the principles of international public law. The issue of power did not disappear, but it seemed to be confined to soft power.

The appearance of conflicts of rules, sometimes with great symbolic significance (industry and environment, trade and social rights, trade and public health, etc.) also underlined the need for arbitration between global goals and national economic choices, in the North as much as in the South. It is this tangled set of concerns, paradoxes and requirements, combined with the ever-present distrust of functionalism and government interventionism that seems to have legitimized the establishment of global governance that coordinates and marks national policies, according to the challenges of globalization. Today, it is difficult to dissociate the term "globalization" from "governance" and "development".

The question of "development" is thus linked to the effectiveness of the system of global governance. The World Bank, IMF, UN and the OECD, in a joint document, defined the roadmap of good policy on an international plan: reducing inequalities in development between countries and reducing poverty in all its forms is the most critical challenge faced by the international community (UN, OECD, WB, IMF, 2000). But it is difficult not to maintain a strong skepticism of the new discourse since, concretely, practices remain the same and policies for opening up to foreign investment and trade remain the panacea for "poor development".

The analysis of global governance should be based on a thorough analysis of changes in the overall legal and institutional framework of competition and accumulation. Globalization and global business strategy have no meaning other than to give it the potential to remove obstacles from making profits. Hence, the importance of a legal framework for the promotion and protection of freedom of entrepreneurship on a global level. The architecture of global governance (on which "development" depends) is based on a consistent set of coercive rules, forms, methods, means of competition and cooperation between economic players whose goal is to organize public and private economic activities globally without apparent discrimination or preferential treatment. These rules may be new (for example, compliance by all countries of free movement of capital or the protection of capital property) or old but, in the context of multilateral agreements, apply to all signatories without discrimination (for example, respect of the most favored nation clause for foreign investors, regardless of their origin). This architecture is global insofar as it assigns an inalienable legal status to economic stakeholders whose activity goes beyond the strict boundaries of a national economy. The organization of cross-border economic activities is only possible if the international firm acquires a legal status, that is to say a full recognition status that confers rights and obligations in any country, provided that those rights and obligations are the same from one country to another. In this context of establishment of supranational rules, it is clear that all countries should review their laws and constitutions to make their legal systems compatible with emerging international laws. Under these circumstances, the only possible road to development remains the capacity of economies and their stakeholders to transform the constraints of globalization into opportunities.

The primary purpose of the power centers that govern global governance is the promotion of the national and international private sector where the opportunities to make globalization a beneficial process to all are supposed to lie.

Wage moderation policy (to keep production costs low and attract international capital), liberalization of capital markets and privatization remain the key words, despite their failure following the implementation of SAPs and the Washington Consensus principles (as discussed above). The fact is that the rules of the global economic game are set, in most cases, in structures that embody a strong asymmetry in decision-making powers; they are the preferred instruments that serve the interests of industrialized countries and the interests of powerful private stakeholders within them, such as financial groups and large multinational firms. In the current situation, it is fundamentally far from this system of global governance whose legitimacy was logically and theoretically based on the need to mitigate the paradoxes created by globalization through reconciling the conflicting interests and objectives of all stakeholders.

At this level of analysis, it would be reasonable to assume that in a globalization process where it is not the system itself, global governance should probably be considered as a new modality of politics for which the objective is less to exceed in complexity but rather to control and stabilize the tensions that are inherent to that complexity. The architectural flaws of global governance relating to "development" raise the question of the leeway that developing countries have in order to assert their claims as a stakeholder. What short, medium and long-term actions should be taken? Should we give the State back its traditional policy tools that proved to be so ineffective during the great inflation of the 1970s (loose fiscal and monetary policies, control of exchange and imports, depreciated currency, etc.)? Should we admit that it was a serious error to advocate neutrality of State actions in the development process, when its role seemed to be more crucial and decisive than ever? Or should we work on the concept of an "astute State" more accurately? The fact is, as we will see later, the new challenges of globalization call on us to reflect on a new development model. If open borders and economic liberalism are the two pillars on which the system of global governance rests, the achievement of this "new development model" should lead economists and political scientists to rethink future economic policies.

1.4. The system of "global governance" under scrutiny

Without supreme regulatory power, globalization and its constraints and challenges may result in unpredictable functioning of the global economy due to conflicts of interest between rival state entities. Fear of economic and political conflict has been used to justify the introduction of a global governance system [MAR 03]. The integration of national economies into globalization is entrusted to international institutions that are deemed to be exempt from state control. Transnationalist theses add to this, demonstrating internal–external continuity and depriving the State of its latest capabilities against the necessary emergence of a supra-State regulation system.

Hidden behind the global governance goals, the problem of relevance of development policy implementation is emerging. Well-defined post-war guidelines (prepare and organize open trade, finance development, conduct proactive industrial policies, etc.) were replaced by a set of goals that do not seem to fulfill a unified vision of the future, even as economic integration (openness to international flows of goods and capital) and the dismantling of State power (economic liberalization, predominance of international treaties) have become an end in itself. This encourages us to see globalization as a process that inevitably leads to the difficulty for States to set goals and develop ways to achieve them: industrial choices, innovation policies, integration of FDI in a diversification program of economic activities, etc.

1.4.1. *Global governance as a substitute for economic voluntarism*

How does the astute State differ from the minimum State and how is it more apt to promote rapid, equitable and environmentally responsible development? In fact, expenditure, operations, administrations, prerogatives, staff, aids to the productive sector, etc., are steadily decreasing. Consolidation and deleveraging remain the primary objectives of economic policies promoted by international institutions. From this perspective, the role of the State in the process of economic development seems weakened. Without any real economic policy instrument (currency being subjected to international rules and fluctuations; regulatory framework being drawn externally), the vast majority of developing (and even middle-income) countries are not masters of their own economy. Their bargaining power with major international companies in terms of technology transfer, employment, and reinvestment of profits or protection of infant industries is reduced. With national control measures of flows of investment or goods becoming obsolete, national governments are unable to control their economy. The problem is therefore a political issue.

Throughout history, in any economy and particularly in developing countries (Latin America, East Asia, China, India, etc.), the launch of major investment programs was accompanied by the implementation of measures to control foreign investment in the sectors of primary resource, energy, transport and communications, defense and security, banking and finance, etc. However, it is true that the results were (and are) questionable. But with the multilateral liberalization of flows, the host country no longer had the ability to guide foreign investment toward sectors that could promote or strengthen national industries and/or control its market. The lack of effective supervision of activities of international companies reduced the spillover effects on local activities, hindered investment, impoverished local production structures and made the economy even more dependent on external resources and more vulnerable to fluctuations in world markets without any control of debt. The example of the General Agreement on Trade in Services (GATS) is revealing the issue. This treaty, which adhered the WTO Member States, was a treaty resulting from the Uruguay Round in 1994 and provided for the liberalization of services in all sectors, except those closely related to the exercise of sovereignty (justice, army, public order, and State administration). In short, health, education, transport, energy supply, etc., were integrated into WTO mechanisms and decisions, and were subjected to market forces. Services' privatization policy, promoted by many governments, met the logic of dismantlement of the welfare State, which was considered to be too expensive.

If we hold to existing theory and studies, the idea of global governance does not exclude the fact that the State may have a social and economic role to play. However, in practice, the weakening of the role of the State in development causes the architecture of global governance to be questioned by those who defend the political framework of the nation-State in the name of a sovereign conception of development. Indeed, today it is difficult to deny that the State is no longer the same unified actor that shares its internal and foreign policy initiative with other international stakeholders and that in many respects, its regulatory function of national economy is fading. Presumably, national interest has become blurred and ambiguous, and rigorous economic policy is giving way to a superior organizational form that itself incorporates singular forms of special interests. This “devaluation" of the power of the State in Southern countries prevents the implementation of proactive economic policies, leaving global governance to take the lead.

Currently, the action of international institutions as leading players in the global governance system is often questioned. Not only the goals, but also the nature of decision-making processes on which this system is based, are questionable. Indeed, the strategy of liberal reforms underpinning the process of globalization is perceived as a gear in which each reform has two objectives: first, to respect a constraint or seize an opportunity for globalization; second, to create a new strain to reduce the State’s leeway.

For example, the role of international financial institutions in the management of the world economy raises many conflicting opinions. While the IMF’s official mission is to ensure stability of the global financial system and the World Bank has a mandate to finance development, these two institutions have come together to play the role of fireman and policeman of the international system by conditioning aid for liberal reforms. However, economic theory has challenged the certainties of traditional models and no longer issues an unequivocal message about the effects of trade liberalization on development. According to standard economic theory, the international division of labor and specialization of national economies in production, for which they have an abundance of capital or labor, are not only beneficial to a particular country but to the whole world. Through the relative price mechanism of goods and factors, costs would drop and people's living standards would improve. The free movement of goods and that of capital is the necessary and sufficient condition for reaching global welfare. But history shows that there are impoverishing and discriminating specializations. It is also an assertion argued by the structuralist approach: it questions modes of insertion into the global economy as the primary and sufficient condition to trigger a sustained and consistent development process. The determining factor for development cannot be external demand for primary products. But historical conditions for development of the world market meant that developing countries were enrolled in the international division of labor as exporters of raw, agricultural and mining materials and as importers of industrial products and/or consumption. If the terms of trade deteriorate (as is regularly the case over a long period), these countries have to borrow in order to finance imports of food and industrial products, compounding their external deficit and, consequently, their external debt.

This reality explains the results of the study by the Philippine sociologist Walden Bello [BEL 02] looking back at thirty years of economic liberalism. According to the study, 80% of loans from the World Bank have benefited a limited number of developing countries to access financial markets. These loans prove, moreover, to be ineffective if we consider that the World Bank itself estimated a 70% failure rate of its projects in poor countries. Indeed, the application of the liberal principles of global governance does not clearly result in better global allocation of production resources. Countries that show the most advanced development and adequate attractiveness policies (see the emerging countries: Brazil, India, China, South Africa, etc.) are those that structurally host the largest volume of foreign investments and have an important place in international trade. Countries with major transport, telecommunications and energy infrastructure, where scientific and technical potential is the richest, and with large solvent markets, etc., are the ones that attract global firms.

The more production and innovation systems are developed on a national basis, the more the economy in question is able to integrate the global logic of the operation of large companies. The liberalization of capital markets, positive interest rates, and the facilities and "national treatment" granted to internationalized companies open new perspectives for financing development. But according to the UN Conference on Development (UNCTAD), in the 1990s and early in this century, 90% of FDI in developing countries have gone to a small group of "emerging economies", against about 50% before the outbreak of the debt crisis (late 1970s). The LDCs received 1%. These countries, unattractive for FDI and dependent on volatile private capital, are forced to contract multilateral loans packaged with reforms causing deeper imbalances; yet they still face another problem of the global governance system, namely the constant fall of official development assistance (ODA).

The virtuous cycle of investment and growth is closely linked to profound changes in the economy and, in particular, to the development and diversification of industry. Even during the debt crisis, East Asian economies continued to rely on industry and high value-added technology-intensive services, through protecting high-technology sectors, rising up the value chain in some sectors (for example, microelectronics) and investing in the development of services (such as banking, insurance and engineering). However, most Latin American countries experienced deindustrialization and African economies went through a "premature deindustrialization" (Table 1.1).

Region	1960	1970	1980	1990	2000	2010	2017 (projection)
Sub-Saharan Africa	15.3	17.8	17.4	14.9	14.9	13.6	13.0
West Asia and North Africa	10.9	12.2	10.1	15.6	14.2	14.6	15.1
Latin America	28.1	26.8	28.2	25.0	17.8	16.8	16.2
South Asia	13.8	14.5	17.4	18.0	15.7	15.1	14.9
East Asia (excluding China)	14.6	20.6	25.4	26.8	27.0	27.4	27.2
China	23.7	30.1	40.6	33.0	34.5	32.3	31.9
Developing countries	21.5	22.3	24.7	24.4	22.7	21.9	20.9
Developed countries	28.9	28.3	24.5	22.1	18.9	17.6	17.2

Table 1.1. *Share of manufacturing in GDP by region, 1960-2017 (in %) Source: [KOZ 04]*

The consequences of systematic deregulation are symptomatic of the constraints and paradoxes in the new world order and of the inability of major stakeholders of global governance to manage it effectively and globally. The first structural responses to financial crises that regularly cross the global economy show, in fact, a return to protectionism in major countries and blocks of Northern countries and simultaneously, a reconfiguration of the financial sphere.

Protectionism is to encourage producers operating in the country itself over others, either through limiting the entry of foreign products into the national territory by quotas or hard-to-reach standards (health, labor, environmental, etc.), or through artificially enhancing the competitiveness of local products by duties on imported goods or subsidies to local producers. According to the World Bank, since the 2008 financial crisis, protectionist measures have been increasing, as well as a rise in anti-dumping measures to prevent the entry of foreign products at excessively low prices. The stimulus packages launched since 2009 (financing of transport, communications and energy infrastructures, bank debt redemption, continued low interest rates, etc.) in major industrialized countries accentuate these protectionist tendencies.

For example, the European Union and the United States routinely resort to protectionism in order to protect and increase the export capacity of their businesses. This is the case for industrial agriculture into which these two protagonists pour a billion dollars per day to support it. This worsens the global nutrition problem due to the low purchasing power of the populations (mostly agrarian) in LDCs. The European Union and the United States demand that their manufactured goods and agricultural products, as well as their service companies, freely penetrate the markets of the world. But at the same time, they are the first to require the protection of intellectual property; 90% of patents are held by Western firms. The limitation of knowledge flow creates a lasting superiority of the North over the South.

However, new financial strategies guide development policy toward the management of debts at the expense of growth. Helped by rating agencies who decide on the reliability of borrowers, financial institutions spot the most fragile States and speculate on government securities. This raises guarantees on State loans (credit default swaps: CDS) and mortgages weak countries in the long-term. However, this situation develops the inventiveness of States with performing economies in terms of protectionism. In an open economy, where finance sanctions the decision and the investment act, the States of industrial and emerging countries use regulations to justify expansion effects or relative decline. These States give up their power on their territories to the private sector and extend their trade policy through international bodies, treaties and standards to the benefit of large companies, which through mergers, acquisitions and equity investments can thus increase their power in the global market.

Therefore, we should learn from previous development efforts to advocate for programs with realistic goals. For example, we should accept, as Joseph E. Stiglitz argues [STI 02], the gradual and differentiated international opening up of developing countries according to their objectives as did (and still do) industrialized countries that have built their economies by protecting key sectors of their industrialization. But the new agreements on direct investment (agreement on trade-related investment measures – TRIMS) signed in the WTO framework prevent developing countries from protecting their industries, either by substituting imports with local production, or by applying measures to increase "local content" in the case of FDI. However, these countries are forced to implement strict legislation on intellectual property protection (agreement on trade-related aspects of intellectual property rights – TRIPS). The abidance to scientific and technical progress essentially achieved in industrialized countries constrains the establishment and development of national innovation systems in developing countries.

1.4.2. *Toward an alternative model of economic growth?*

Global governance formalizes a commercial, productive and financial framework drawn by the political choices of major economies [UZU 10b], while promising to developing countries to accelerate their industrialization thanks to free trade [UZU 05, UZU 10d]. But according to J.K. Galbraith [GAL 84], the industrial world applies an economic model in companies that doesn't take historical processes into account: the great powers apply standard development programs without considering the historical characteristics of less developed economies. These programs express the state of the economy of these powers and leave little space for sociopolitical conditions on which capital formation could be based. For Galbraith, and according to the experiences of former industrialized countries, the prerequisite for economic development is political development, itself correlated with the democratization of education. A political system must be stable and predictable, honest and efficient; citizens (educated and informed) should be the stakeholders. But in order to achieve this, citizens must be educated. Good governance starts with the organization of a system of basic teaching, education and training of individuals.

Education is, indeed, the foundation of political organization from which the process of development emerges. World Bank reports share this view. Free, compulsory and good education will break the culture of poverty. It is also closely connected to the participation of individuals in decision-making in economic and political fields. A good general education sets the stage for more specialized education in a technical, scientific or administrative field. Itself forming the "human capital" that is essential for the selection, design and/or absorption, utilization and development of necessary technologies that are compatible with the economic development project. General and specialized education is also involved in the formation of a stable political system that is able to give meaning to development and provide the material, financial, cognitive and institutional resources needed to achieve it.

In an open economy, what direction will lead to development? What are the conditions and what types of institutional tools can be used to stabilize the economy, to control the flows and the stocks? International institutions have realized that without a State, in the absence of a representative and legitimate political system, development options and managerial choices are limited. Table 1.2 shows, on one side, the impasses that the implementation of proposed measures in developing countries led to (especially for the most fragile economies) since the debt crisis (late 1970s); on the other side, it shows some institutional arrangements to escape underdevelopment.

Global governance and development crisis	Institutional renewal and economic organization
– Instability and political crises – Unemployment, poverty, increasing social inequality – Deficient markets, informal practices – Financial and regulatory institutions in their infancy – Neglected collective infrastructure – Economy subject to the hazards of the international environment – Fragility due to unpredictable capital movements	– Promotion of a predictable political system and rehabilitation of the role of the State – Priority to education and collective social infrastructure – Coordination system of market stakeholders and decision-making capacity of the State – National production resource control procedures (capital formation, income taxes, currency) – Centralization of a domestic savings system – Differentiated international opening according to national goals

Table 1.2. *Development, global governance and institutional renewal*

Table 1.2 shows the need for developing countries to create and strengthen the systemic relationships within their economies. Few countries that qualified as developing countries at the height of the political economy of development (1960–1970) followed the path opened since the 18th Century by the countries qualified today as industrialized countries, to attain the magic triangle: growth by opening new markets to satisfy the greater needs; establishment of a national economy system capable of ensuring an endogenous process of capital accumulation through innovation, mastering the financial circuits, investing and selectively opening to international trade; emergence of an autonomous political process for defining the national economic development project. The current architecture of global governance does not allow this type of economic intervention, except for so-called large emerging countries that are endowed with primary resources, a sufficiently large market to launch industrialization and stakeholders (State, businesses and entrepreneurs) who are interacting with and having sufficiently important common interests to develop particular modes of regulation of the national economy.

Research on the "third world" has long since reached the conclusion that most developing countries are political entities born from external constraints and do not reflect their current (or past) social and economic structures. Most often, their political systems are largely imported and reflect a projected socio-economic status that they can only aspire to. Development economists know this and forcefully point out: from the moment where a national rise is without the footprint of the political system on the model, the structure and the pace of development, any economic policy comes up short against the reality of structures. Under these conditions, the formation and consolidation of a national economy are priorities. Mastering accumulation means mastering the market, controlling natural and production resources, launching procedures for regulating and reforming the economy.

Historical experience of industrial and currently emerging countries indeed shows that the constitution of the national economy, the model, the institutions, the structure and pace of economic development are largely determined by the political system, and not the opposite. Economic development policy is thus subject to six conditions: understanding of national capacities; ability to mobilize and strengthen them; definition of objectives to be reached; identification of bottlenecks; choice of technology; reformist capacity of the State. The representation of the market economy is based on "private effort to accumulate capital" but the neutrality of currency and the State in classical and neoclassical, Smithian and neo-Smithian models distorts the analysis of social organization and its reformist and adaptive dynamics. History, however, confirms the most basic realities and reveals in correct terms the fundamental problems of development: money creation, fiduciary revolution, State, system training and predominance of individual national systems. In globalization, the big picture is how to formulate a policy and how to develop the tools (commercial, financial, regulatory, scientific and technical) to design and implement development projects.

The current rules of global governance are born and applied in the context of a world economy composed of power centers that are unequal in size and power, centers that are structured more or less solidly but maintaining asymmetrical relations. They also attribute a status to global companies and freeze positions in the global trading system. Hence also a renewed interest in "development economics" and the implementation of active policies, starting with the formation of a relatively independent national credit system in relation to international financial flows that ensure the capacity of money creation and the control money circulation. In turn, the effectiveness and efficiency of the financial system depend on the control of the labor market and the evolution of employment. The latter being dependent on the distribution patterns and allocation of surplus, itself dependent on the control of the market, natural resources or technologies.

After the Second World War, countries seeking to develop were encouraged not to develop their own capacity to innovate but rather to import the most advanced technologies as heavily as possible. This mimicry largely explains the failure of development strategies. It induced debt crises, greater dependence on Northern countries, a brain drain, poverty and an explosion of migration. The idea that we can simply import foreign technologies without appropriating at least part of their production conditions is unfounded. The demands of an active integration into the world economy combined with those of the structuring of a competitive production system lead the researcher to question the capacity of developing countries to develop their own innovation systems.

The issue of development is therefore to make *globalization profitable* and from this perspective, it is clear that there is no alternative but to renew reflection on the relevance of a new and proactive role of States and more particularly, in science and technology policy and innovation policy. The merits of such policies depend on the existence of so-called market failures. Institutional, structural, productive, distributive, financial, etc., inefficiencies are all market failures that legitimize public measures to deal with them, with a view to ensuring effective operation of the production system and promoting its integration into the global economy, according to development requirements. But initially, economic policy should move toward the establishment of institutions guaranteeing: (1) the best possible allocation of resources in order to avoid dominant market positions; (2) the economy stabilization to effectively and fully use these capital and labor resources; (3) the social transfers to meet the basic needs of the population (food, health, education and housing); (4) the financing through fiscal policy for the production of public goods.

Unlike analyses that reduce the State to an agent that creates distortions and/or collects income, new development macroeconomics connect growth, competitiveness and financial balances and introduce structuralist elements in a macroeconomic framework [UZU 10e]. Thus, in the context of a globalized economy, the "pro" State (promoter, prospector, protector and producer) becomes a central agent of development that aims to change the modes of integration of the country in the international division of labor (IDL). This requires a change in the nature of specializations through the implementation of industrial and innovation economic policies, coupled with institutional and social modernization, which act on the productive structures. By this logic, UNCTAD proposes the establishment of development policies whose objective is to stimulate and monitor structural transformations.

The revival of the State is concomitant to a change in its modes of action in fundamentally altered national and international contexts. The design of an alternative development model assumes that national States have significant enough "leeway" to control its economy. This thesis will certainly go against the Washington Consensus (which is based on the objective of convergence of policies for undifferentiated integration in the global market), but it does not reject the relevance of active and phased integration policies in international flows of goods and capital.

The economic policy of development in globalization involves the definition of strategic objectives at the center of which is the design and promotion of innovation capacities and the formation of an innovation system capable of capturing and producing knowledge; the national economy thus can take advantage of technological advances and simultaneously contribute to their achievement. The performance of a country's innovation system determines the structural competitiveness of its economy since micro- and macroeconomic performance depends on "non-cost competitiveness". In other words, economic development cannot be based on the exploitation of static comparative advantages derived from factor endowments, but rather on building competitive advantages from the implementation of intensive production processes in scientific and technical knowledge and technologies. While technological and social innovation becomes the basis of development, the priority is given to incentives for R&D and human capital that guide the overall economic policy; financial considerations of development then move to the background.

2

Innovative Capacities and Systems of the South in Globalization

The challenges of globalization, beyond the system of global governance, demonstrate the importance of innovation and technologies as tools for growth and sustainable development in the North as much as in the South.

There has been widespread research on innovation in Southern economies over the last 30 years [KIM 97, HOB 95, AMS 01, BEL 97]. More recently, the issue has been whether the concept of an "innovation system", as developed in the 1980s, is a useful tool for studying growth and development [LUN 09, MUC 03]. The idea is to use this concept to understand learning and innovation processes in developing economies. Therefore, it would not only serve as a tool for comparing the technological performance of countries, but also as a promotional tool for economic development.

Initially applied in Northern economies, the concept of "innovation system" was introduced into development economics for the construction of a set of economic policy proposals that combined the demands of globalization with the specificities of Southern economies [ALT 09]. The structure of the private sector in developing countries and its performance differ largely from Northern economies. This requires considering, among other things, a number of key elements such as weak legal systems (less secure property rights and higher transaction costs), differences in demand conditions (low purchasing power, demand for more or less sophisticated products), poor infrastructure (high transport and production costs), weak education systems (from primary school to universities), price instability and greater capital volatility.

The aim of this chapter is twofold. First, we attempt to understand how to examine systemic innovation in the South and then we discuss the difficulties of achieving it. The concept of the innovation system will find its meaning in a broader sense, given the nature of innovation and especially the connection with building necessary innovation capabilities. But for this instrument to be consistent and applicable, it is necessary to understand on what strategies innovation of the South is based and how the North/South technology transfer is ensured, as the technology comes from innovation systems that are richly endowed with scientific and technical resources. Thus, it is important to create national learning capacities but these must be accompanied by internal learning processes that are based on endogenous knowledge. However, institutional and innovation policy failures prevent many developing countries from building skills and consequently create systemic links between potential and current innovators and between endogenous and imported innovation.

2.1. Innovation for economic development

An innovation system describes the relations between institutions (scientific, technological, industrial, commercial, financial and political) both private and public (companies, research laboratories and engineering entities, administration, etc.). These relations are often formed through financial and information flows and movement of skilled and qualified people. The purpose of such a system is to produce innovations (new organizations, new goods and processes, new resources and new combinations of productive resources). These systems are national (or local) due to the specific features of their organization to which legal public regulation systems apply.

Industrial and innovation policy is dedicated to the establishment of arrangements that are necessary for industrial and economic development through the strengthening of links between competitiveness and innovation. These links are assessed by the ability of countries to produce and export manufactured and technological products and the structural indicators of industrial performances: skills, R&D effort, direct investment from abroad, paid royalties and license payments abroad, modern infrastructure, etc.

Disparities in the level of industrial development between developing countries are significant. We can observe a high concentration of technological capabilities in a few countries (so-called "emerging" countries) and a worsening of the situation of the poorest countries. To improve their position in global competitiveness, various experiences show that developing economies need to selectively open to international trade and foreign investment while supporting this opening policy by building their capacity for innovation. Integration into global value chains is one way to access the highest technology and to encourage, through learning, their mastery.

But should industrialization not take into account the fact that innovation does not have the same meaning and the same form in all economies? Recent history shows that the recipes of the North often failed in the South. The interest shown in the nature of the innovation process (where collaboration is crucial) and in the varied forms of innovation (organizational, social, etc.) of the so-called "lagging" countries could perhaps form new sources of wealth and competitiveness.

2.1.1. *Understanding globalization through technology transfer*

If the use and dissemination of modern technology, which diversifies supply and increases labor productivity, is faster in developed countries than in developing countries, then the current technology gap is widening. Since it is foolish to argue that the South must reinvent everything, it is obvious that technology transfer and their adaptation to local sociotechnical and economic contexts become priorities. Technology transfer cannot, as a matter of fact, replace the existence of R&D and innovation capacities in a country.

Theories on technology transfer underwent a wide expansion in the 1980s. They were related to the intensity of globalization and the increase in the international competition. This transfer was a crucial factor for economic and social development. The economic catch-up assumed massive importation of technology from industrial countries. Technology transfer was defined in UN regulations as a transfer of systematic knowledge for the manufacture of a product or for delivering a service. It was a transmission of knowledge between companies and public institutions belonging to different countries and an exchange of equipment and technology between countries. Technology transfer can take place from one firm to another firm in a foreign country, from a research laboratory to a newly created firm or one that already exists. The innovation process in developing economies comes from a process of this technology transfer. Here, "technology transfer" means the dissemination and the capacity of developing economies to capture foreign technology in order to integrate it into their production operations to launch products and competitive services on (local and/or global) markets and increasing added value.

Although there was an abundant literature on technology transfer in the 1980s, Hendrickx [HEN 96] sums up the three existing categories: material transfer, transfer of concept and transfer of capacities:

1) Material transfer is treated as the simple import of new equipment and related techniques. Local adaptation is designed as trial and error. The core technology is trapped inside the physical goods;

2) Transfer of concept is achieved through transfer of certain plans, formulas or books in order to make a domestic product. This is a simple transfer of information, data and guidelines for creating basic skills;

3) Transfer of capacities is in turn directed toward a transfer of scientific knowledge for the production of locally adapted technologies. This transfer leads to the development of our own technologies from imported technology.

Teece [TEE 76] clarified two main forms: a physical form with the characteristics of various products and equipment and another more informational form, which may be acquired if the equipment is actually used. Technology transfer should include four distinct elements, namely one content, two parties and a result. The content consists of a transfer of tangible and intangible assets for operating the transferred technology. The two parties are the exporter and importer of the technology. The result is relative to the recipient's ability to operate the imported technology. UNCTAD classifies three forms of technology transfer: its production through multinational companies and R&D, its transfer through multiple forms of external growth strategies (mergers and acquisitions, franchising, license acquisition, technical assistance) and its diffusion through the generated externalities. For Berthe [BER 97], there were three different categories of technology transfer: appropriate technology relevant to the needs of the local population; advanced technology associated with the use of technologies from developed countries; technology recovery to recover and transform technology from developed countries.

But this technology transfer also came at a cost. Teece [TEE 76] described four components in particular: the costs of technological engineering exchange related to the theoretical basis, the engineering costs associated with the transfer itself, the R&D costs and the training costs. Technology transfer was much influenced by the efficiency of the process of internationalization of firms as a basis for multinational companies and direct foreign investment [BUC 76].

Technology transfer can take place in various forms, such as:

– intellectual property rights license agreement;

– know-how transmission agreement;

– research and license contracts;

– technical assistance and vocational training contract;

– contract for realization of industrial complex;

– direct investment controller joint-venture;

– mixed contract including one or more forms.

But here, we are less interested in the forms than in the processes related to their development. Although the literature is abundant, we can nevertheless present two major theoretical orientations: that of the technology gap between rich and poor countries and more recently, those related to capacity and skill building. The latter literature insists on the ability of developing countries to internalize foreign knowledge. We can consider that developing countries are not mere imitators of foreign technology and can appropriate foreign knowledge by creating a capacity to innovate.

The technological gap theory is far from recent. It was developed by Posner in 1961 to highlight the strong competition of the economies between high and medium technology sectors [POS 61]. By focusing on innovation, certain countries gained a technological edge in a sector, which gives them an export monopoly in this sector, regardless of their advantages in factor endowment. A technology trade gap arises if foreign consumers express a demand for new products that require a certain period. This imitation period will be ever longer as the innovator maintains a cost advantage that is often associated with economies of scale arising from the existence of an extending market. New products appear and others disappear at a high rate that corresponds to rapid technological change.

The works by Vernon [VER 66] were in line with this neo-technological approach. Focusing on the product lifecycle problems, Vernon quickly extended his analysis to international trade. According to this author, any innovative product undergoes a lifecycle that has four phases: introduction, growth, saturation and decline. These phases can be coupled with the technology dissemination procedures on foreign territory:

– The launch phase corresponds to a diffusion of the product in the domestic market with high incomes and strong demand for innovative products. The firm behind the new product may have a temporary monopoly.

– The growth phase is marked by widespread nature of the product which becomes a commonplace on its home market, encouraging the innovative firm to export the product to similar markets.

– The saturation phase is characterized by the fact that the diffusion of the product is no longer progressing on the domestic market. This is also the time when the product is imitated by other firms. The innovative firm conducts technological transfer to adapt and sell its product in developing countries.

– The decline phase corresponds to the replacement of the product by a more innovative product: lower consumption and the production of the good is completely relocated in order to benefit from lower production costs; competition is on the price and not the innovative nature of the product.

Although this thesis has the merit of describing the gap between rich and poor countries, it says nothing about the means in order to reabsorb it. This thesis considers that it is the rich countries that can innovate and conquer foreign markets; this is causing the delay of other countries. That being said, a country that imports technologies and goods should be prepared by adapting its human and technical capital toward global production (and consumption) requirements. The success of the technology transfer depends on the recipient country's capacity to assimilate the imported technological change. The access to foreign technology should be correlated with the internalization of knowledge that it contains. Therefore, one must refer to the literature on the capability and learning of firms to better understand the dynamics of technology transfer. Importing countries cannot just be the passive stakeholders of technology transfer. They must create a favorable technological environment encompassing all the factors necessary for the acquisition, adaptation and efficient operation of the imported technology [KAT 82]. But the exporting country should also ensure adaptation of the technology to local conditions of developing countries.

Technology transfer is essential for developing countries: if successful, it can provide them with a competitive advantage. This transfer, linked to the capacity of technology diffusion from the North to the South, has often been associated with the simple accompanying measures to improve technology mastery. The failure of the major technological and industrial projects in the period of 1960–1980 also showed that this transfer could be limited to a perfect imitation of technology. Therefore, work in this area was more oriented toward the diffusion of technology coupled with its actual creation. In this case, developing countries had to master the technological knowledge, going beyond the mere ability to copy or imitate. It is now widely believed that developing countries must develop their own skills from their absorptive capacities of imported knowledge. This capacity results from both macroeconomic factors (education policy, R&D, etc.) and from internal learning processes. For these countries, the aim is to combine the assimilation of external technology with their own capabilities/resources in order to launch a new technological trajectory. A valuable instrument to encourage this dissemination and creation of technology is the building of innovation systems.

2.1.2. *Innovation for development*

The issue of inequality and poverty is essential in the strategies for acquiring foreign technology and innovation of the South [COZ 09]. These problems are also rooted in the socio-economic situation of these countries. But innovation can only occur in an appropriate development context. To link development with innovation is important because the two interact positively. As pertinently emphasized by Juma *et al.* [JUM 01, p. 649], "many public debates are driven by negative remarks about science and technology and the result is that there are few investigations on how to shape technology to cope with the new challenges of development".

The innovation system is an analytical tool for understanding growth and economic development, but also the performance of national economies within globalization [LUN 02b]. Two aspects must be considered before studying the relationship between an innovation system and economic development: the introduction of the concept of innovation in development theories and the prospect of economic development within the systemic approach to innovation. The first aspect refers directly to the lack of consideration of innovation capabilities in the development models and strategies [GRE 01]. It is undeniable that the new endogenous growth theories helped to introduce technological change and human capital formation in growth models and consolidated the role of technology in economic growth. However, they remain limited in systemic innovation design, which is rooted in evolutionary and institutionalist tradition. We therefore speaks of missing capability to justify the need in order to create a new entryway through innovation and learning in development strategies [JOH 03]. In this sense, technological capacity becomes an essential component of development [JUM 01].

On the other hand, although the concept of innovation has not exactly been introduced into development theories, a systemic approach to innovation has meanwhile been particularly influenced by developmentalist theses. Development theories became the new fields of application for the concept of systemic innovation [LUN 02b]. While the national innovation systems approach draws on the concept of systemic feedback from Hirschman [HIR 74] and Steward [STE 77], it borrows the ideas of positive and negative feedbacks, "cumulative circular causation" and vicious or virtuous circles from Myrdal [MYR 57]. It is also from Hirschman [HIR 81], in his famous essay "*Essays in Trespassing: Economics to Politics and Beyond*", that we discover the interdisciplinary nature of the approach [EDQ 97]. Interest in the interactive processes between innovation stakeholders also comes from Hirschman's writings [HIR 81, p. 75]: "Development is essentially the story of how things lead to new things, and from a specific point of view, the links represent this story". Ultimately, the main ideas of development economics are re-exported in the works on innovation systems.

Reducing poverty and improving income distribution remain related to the capacity of developing countries to use, disseminate and create knowledge. Economic development is understood through the accumulation and use of tangible and intangible resources (Table 2.1).

	Easily reproducible resources	Resources that are difficult to reproduce
Tangible resources	Productive capital	Natural capital
Intangible resources	Intellectual capital	Social capital

Table 2.1. *Economic development by accumulation and use of tangible and intangible resources (source: adapted from LUN 02b)*

Benefits derived from innovation go beyond their contribution to production. Sen [SEN 03, p. 385] noted that "to recognize the role of human qualities in economic growth... is no indication in any way as to why we need to give as much importance to economic growth. However, when our concerns are, ultimately, the freedom to live the kinds of life that people have reason to hope for, then the role of economic growth – because it contributes to the creation of multiple new opportunities – can be integrated into a more fundamental understanding of development as a human capacity expansion process to live richer and freer lives". Since we cannot reduce human beings to a means for production and that they also have a real purpose, innovation systems must also consider such "multiple new opportunities" through the workings of economic development.

The idea is to understand how learning and innovation enable the capturing of "multiple new opportunities" by various economic stakeholders: individuals, companies, socio-economic networks. Measures taken by developing countries to promote scientific research and technological innovation are key elements for economic development [JUM 01]. Although Lundvall *et al.* [LUN 02b] demonstrated the relevance of combining learning/development in Denmark, the issue in developing countries was raised as follows: "there are reasons why we need to rethink the concept of innovation system in different contexts to make it applicable to African States (...). We need to take living conditions into account and their links with learning and innovation" [JOH 03, p. 24]. The objective is to know how innovation and learning have effectively contributed to the development of poor countries [DOL 04]. However, although research on innovation in developing countries has developed the concept of sustainable development as an essential and new framework to consider (such as ecological values, the influence of social capital on the processes of learning or the problems of uneven development) [SU 04, SEG 03], it also emphasizes the need to coordinate internationally with major trends in research programming and with the recommendations of international organizations.

In this sense, Narayan and Petecsh [NAR 02] clearly showed how listening to poor countries is imperative for connecting innovation and development of less developed countries. Again, the viewpoints of international institutions and the World Bank are open to criticism because it is not about listening to the voice of Western countries but rather to listen to the voice that poor countries would have wanted them to hear [DOL 04]. Innovation has very contradictory effects on growth and development. Lundvall and Borras [LUN 97] recalled it perfectly with the negative impact of innovation on labor in industrial economies. Competition requires the provision of means to promote increased business competitiveness. But innovation is no longer limited to technical innovation; it concerns all individuals, whether as consumers or producers.

Also in this perspective, innovation can reinforce inequalities, since changing technologies reinforce income distribution gaps. Technological change exerts considerable pressure on wages and employment through competition faced by workers and job losses as the corresponding production is replaced by imports. Indirectly, it is done by stimulating innovation in rich countries in order to offset their higher labor costs through higher labor productivity and thus compete with poor countries through the quality and variety of their products.

Conversely, development must also be a valve for innovation. Yet in many developing countries, inequality is hindering the innovation processes by blocking sources of learning (access to higher education, quality of education, importance of the health system, communications infrastructure, etc.). The existence of strong inequalities in the economies of the South undeniably hinders the learning and innovation capacity of economic stakeholders. Moreover, the innovation systems of the South describe social situations in which conflict is ever present: conflicts within the innovation system (institutional capacity problems, inter-institutional problems) and conflict on the macro-social level (education systems, legal systems, etc.) [ARO 03].

It would therefore be difficult to believe that innovation in developing countries can *de facto* contribute to improvement in the quality of the productive fabric in Southern economies. Especially that, as we have seen, these countries suffer from a lack of market and social links, “bad governance”, institutional deficit, low efficiency of public policies, degradation of the environment, etc. We cannot be optimistic about a dynamic and coherent relationship between learning, innovation and development [JOH 03].

Nevertheless, if we look more closely, the reality in developing countries is less inclined toward pessimism than some analyses lead us to believe. Indeed, these countries are richly endowed with natural resources, in know-how as well as specific resources that require the activation strategies of stakeholder (entrepreneurs, companies, non-profit associations, public education, research and engineering institutions). These environments contain an amount of tacit knowledge that could be the basis of a virtuous innovation process. This knowledge is not immediately recoverable. This gives the involved countries a major role in the exploitation of its specific resources that are tacit knowledge.

2.2. Innovation systems and integration into the world economy

The approach to growth and development based on innovation systems is not originally linked to globalization, but rather aims to identify the peculiarities of national innovation in order to explain innovation performance differences between economies of different countries.

The NIS are defined as the set of institutions that are involved in the production, commercialization and distribution of new or improved products, new processes and services. They also include structures of incentives and skills that influence the speed and direction of technical change. Components of the NIS can be classified into three broad categories: (1) innovative firms and private research institutions; (2) universities, public research institutions and organizations; (3) financial institutions, technology intermediaries, governments and public authorities. These innovation stakeholders interact between them and with technology users and final consumers of goods and services. Thus, beyond the competence of national innovation agents, interactions between these innovators and between these innovators and the technological and sociotechnical environment are central to the analysis of a national (global or sectorial) innovation system.

The interactivity between NIS stakeholders shows the importance of financial institutions, of the vectors of scientific and technical knowledge that are engineering companies and administrations, whose role is to coordinate, promote and guide the NIS through incentives and a legal framework. But the focus on improving interactions within the NIS does not mean that more traditional functions of government are neglected. These particularly include capacity building in public research centers, ensuring an appropriate level of investment in R&D and improving the business climate to attract international investments that bring in and diffuse innovation.

2.2.1. *Innovation capacity and learning process*

Technology cannot be reduced to perfectly transmitted information. It takes a codified and uncodified dimension. We must then consider a view of technology transfer in terms of capacity and resources. The substance of firms will build on what they are capable of doing. Knowledge and skills become strategic assets and play a decisive role in the development of organizations. Competence-based theories recognize the important role of tacit and social knowledge in economic organization. The skills therefore appear as the basis of the competitive capacity of a firm in a particular activity. We will consequently be less interested in the technological gap between rich and poor countries, but rather in the ability of developing countries to use and adapt existing technologies at competitive costs and quality, in line with international standards [LAL 02]. Here, innovation is no longer situated at the frontier of technology but includes catch-up strategies implemented by the followers. Innovation is therefore a dynamic process.

Neoclassical economists have been very interested in allocation problems in a general equilibrium context. The individual agents should make rational choices among the various proposed alternatives according to their preferences and information including stocks of technical knowledge. In this regard, the normative conclusions on the organization of the economic system are known. That said, in its own foundations, this vision could not attain an objective to understand the current phenomena in economic development. Indeed, if firms or economies focused on the allocation of existing resources (capital and labor) and if each separate unit made the same product with the same technique, these firms or economies would become markedly less competitive because of consumer behavior. We therefore recognize that the success of innovation, including all tangible and intangible assets, is more important than the view of optimal resources allocation, particularly in a context where the speed of change is constantly accelerating. This is less about distributing resources relating to labor and capital than about creating and using knowledge through various learning processes. Also, the information and specific knowledge that agents have is less important than their capacities to learn. These include the ability to do new projects, deal with new situations and gain access to new information. In short, companies must constantly seek new knowledge to use in production in order to introduce new products or processes and to improve the existing ones.

Innovative capabilities are built upon the technological knowledge mobilized to assimilate, adapt and modify existing technologies in order to adapt them to a local socio-techno-economic context. Innovation capabilities refer to national efforts to integrate imported technology into local production processes. These are dynamic resources, unlike the production capacity which refers to the equipment and labor needed to produce industrial goods with given efficiency levels [BEL 97]. In reality, two aspects are particularly important: first, all endogenous resources and knowledge accumulated through traditional learning process and second, the effective capacity to absorb and adapt foreign technology [KIM 97].

Learning processes are affected by the absorptive capacity of firms. This is a function of the capability of firms to combine two main elements: the existing basic knowledge and the intensity of efforts [COH 89]. The existing basic knowledge is a key platform for technological learning as it directly influences the learning process and the creation of future knowledge. As for the intensity of effort, it is related to the amount of energy expended by organizational stakeholders in order to solve problems created by new imported technologies. These two aspects are complementary because the learning process in which foreign technology is assimilated largely depends on the scope and quality of existing resources and knowledge.

The arrival of foreign technology cannot exclude the endogenous development of existing technologies. This includes the combination of local technologies with imported technologies. Moreover, technological learning of a company does not develop in isolation. The learning process is full of externalities and interrelationships, particularly those between producers and users [LUN 88]. Few of these links pass through the market; instead they tend to occur informally through rote learning. Also, "national absorptive capacity" [NAR 04] or "domestic technological capabilities" [LAL 00] exceed the sum of individual capacities of firms in the studied country. Reciprocally, the capacity to absorb foreign technology is not only a function of the presence of the related firms in the economy. During the period of learning and absorption of these technologies, other economic stakeholders (related or ancillary companies, public institutions, universities) enrich the acquired knowledge while taking advantage of externalities produced by the efforts (investments) made by individual companies for the successful technology transfer in question. The ability to innovate is thus a collective venture carried and orchestrated by the innovation system.

2.2.2. *About national innovation systems*

The National Innovation System (NIS), a broad concept of which the meaning has largely evolved into a concept that is related to the economic development of developing countries, was formalized in practice and intellectually in the 1990s. Today, it is widely used as an instrument of technological and innovation policies. Although the analytical structure initially rested on industrial countries, it has gradually imposed itself in developing countries.

The definition of NIS differs depending on the authors. We distinguish between a broad approach related to the learning process of economic stakeholders and a restricted approach that is directly related to science and technology (Table 2.2). In developing countries, the idea is mainly to retain a broad approach that considers "the elements and relationships that interact in the production, dissemination and use of new knowledge" [LUN 92].

	Narrow	Large
Formal	Organizations, formal institutions and networks related to science and technology	Organizations promoting innovation in general (technological, managerial, economic and social) formal institutions and networks
Informal	Informal networks and institutions related to science and technology	Informal institutions that influence innovation and informal networks with historical and cultural values

Table 2.2. *Interaction between the formal and informal elements of the innovation system (source: adapted from [SCH 99])*

This choice is related to various characteristics of innovation in the South, as discussed below:

1) Innovation is first and foremost about relationship. It is supported by substantial social capital (consisting of non-quantifiable interpersonal relationships that convey critical knowledge). Although Lundvall [LUN 02a] showed the importance of social networks in Northern economies, these networks deserved special attention in the South [ARO 03]. The impact of social ties on innovation and learning is important since the intensity of these links provides rich sources of (technological and organizational) information and knowledge that are needed for innovation activities. Kuada [KUA 03] showed, for example, how in Africa, social ties promote interpersonal trust. The latter has a positive effect on social relations and on motivation to develop learning in a common action. Social ties through trust, empathy, emotional relationships, are all qualities that allow a faster acquisition of knowledge and the promotion of innovation since they reduce the risk in business, reassuring the innovator and providing new knowledge. In his study of industrial clusters, McCormick [MCC 99] associated the importance of African collectivism with its influences on learning. The cluster effect related to geographical proximity and strong social capital encouraged sharing of innovation and increased the opportunities for learning new techniques.

2) Innovation is as much present in the sectors of low and medium technologies as in high-tech sectors intensive in R&D investment [JOH 03]. While innovations in high-tech sectors are sophisticated and based on the immediate application of science to industry, innovation in developing economies is most often the result of routine learning processes within traditional small structures. Innovation is mainly conducted informally through learning by doing

procedures, through usage and through the interaction between national and international stakeholders, whether formal or informal. R&D activities are not clearly and formally found in entrepreneurial or sectorial strategies. In studies on systemic innovation in developing countries, taking the diversity of the learning process into account is more necessary than in studies of developed world. A large part of learning, particularly through the development of tacit and localized knowledge, refers to on-the-job learning by working directly on the machine or, more generally, by performing an operation [JEN 07].

3) Innovation is strongly marked by the characteristics of the global environment. Innovation in developing countries is often complex and particular. The economies are unstable, vulnerable and innovation strategies are linked to macroeconomic, political, institutional and financial environments [CAS 03]. We talk about an *inert national system* [HOB 95] to highlight the social, political and technical problems encountered by companies in these countries and which are considered as real obstacles to innovation. It is therefore important to understand the history and operation of such developing countries in order to assess their innovation strategies. The heterogeneity of development trajectories and the varied and divergent dynamics of catching up through the development of more or less successful technological capabilities define the particularities and performances of the NIS of the South [MYT 03].

4) The innovation strategies of the South are interpreted through their construction and not by their actual technological results. The innovation systems of the South have the characteristic of being incomplete or unbalanced in the sense that certain aspects are inefficient or simply non-existent [NAR 04]. According to the studied economies, there

are many organizational and institutional rigidities, inadequate macroeconomic policies, low investment in education and R&D, as well as low production capacity and technical assimilation. The innovation systems of the South are generally depicted through their various systemic failures. This makes it difficult to talk about the *a priori* (*ex post*) character of innovation, which is used to describe the process of innovation in industrialized countries with a strong institutional basis and advanced infrastructure. The innovation strategies of the South should instead be of *a posteriori* (*ex ante*) in nature, so that analysis of the learning process can move further upstream, that is to say along with the conditions of emergence and development of the learning process. We generally refer to the terms of *construction* or *promotion* of (national or sectorial) innovation systems of Southern countries and not the terms of *intensity* and *performance* [LUN 02a].

5) Finally, innovation is linked to development strategies and poverty reduction. In developing countries, the unequal treatment of minorities, weaknesses in the system of education and vocational training, the brain drain or poor local governance through corruption problems divert human and social capital building and automatically limit the learning capacity of the society. The rentier economy and the political instability are not conducive to knowledge creation.

The individual capabilities of economic stakeholders depend on the institutional support of the society and its social, political and economic arrangements. The term *missing capacity* is evoked to justify the necessity of creating a new input channel through the learning capacity in development strategies [JOH 03b]. In this sense, technological capacity becomes an essential component of development [JUM 01]. As such, Arocena and Sutz [ARO 03] summarized the spirit of the National System of the South in comparison with that of the North (Table 2.3).

NIS in the North	NIS in the South
NIS is a concept...	
Ex post, developed from the analysis of actual existing processes.	*Ex ante*, mainly used to analyze the virtual nature of the process.
It is a relationship-oriented concept but...	
Innovation relationships are embedded in a dense social fabric.	The network of innovative relationships is fragmented and often more informal than formal.
The concept of NIS is setting the direction of policy design that...	
– Must support the deliberate efforts to enhance dialogue between key players of innovation. – Reflect the (historically constructed) social legitimacy of national efforts in order to cause the construction of knowledge and innovation.	– Are reduced to a "copy" of institutions and do not generate specific actions that promote existing innovation, which are informal and "interstitial" in nature. – Reflect the lack of consensus on the role of R&D, the consequence of a misdiagnosis between adapting foreign knowledge and creating own knowledge.

Table 2.3. *Comparison between NIS of the North and of the South (source: adapted from [ARO 03])*

According to an UNCTAD report (2012), some developing countries accelerated their economic growth to improve their innovation capabilities (see Box 2.1 for the example of Taiwan). In countries where dense innovation networks were created between local and foreign firms and between local businesses, the national context, by fostering innovation, has helped to strengthen their technological capabilities.

Taiwan is one of the four Asian tigers – four Asian countries or regions that recorded an exceptional growth rate in the 1980s and became newly industrialized countries in the 1990s: South Korea, Hong Kong, Singapore and Taiwan. The evolution of its national innovation system demonstrates the relationship between the two concepts for the definition of NIS: strictly technological and extended to social capital.

The Taiwanese innovation system evolved in two phases: the catch-up in the period of industrialization and the development of endogenous capacity for innovation. In the first phase of upgrading and industrialization, the Taiwanese industry benefited from the integration into the international labor division by becoming the supplier of multinationals in favor of development (1960–1990). The Taiwanese NIS was then created *de facto* by the government in order to facilitate the acquisition of foreign technology (through the purchase of means of production, FDI, joint ventures, licensing agreements, etc.) and to build local absorptive capacity (in particular by investing in training to meet the requirements for qualified personnel and thanks to support for the qualified Taiwanese returnees implemented by the government, see [SAX 01, TSA 03]). During this period, under the leadership of the government (promotion of high-tech industries by academic spin-off and networking of stakeholders through the science park construction policy), Taiwanese companies were specialized in high value-added industries, such as electronics and Original Design Manufacturer and Original Equipment Manufacturer (ODM/OEM) [AMS 03].

Voluntary policies have allowed Taiwan's industry to acquire and accumulate new knowledge, develop knowledge capital of firms and develop high-tech sectors. However, the Asian financial crisis in 1997 and the bursting of the Internet bubble highlighted the weakness of the Taiwanese national innovation system: lack of innovative capacity of Taiwanese companies measured by the number of patents and weak public research as measured by academic publications.

Meanwhile, confronting the international competition exacerbated by globalization, including competition from China in the late 1990s, the government's new policy sought to strengthen and diversify the internal links of the country's NIS. The new approach aimed to strengthen the flow of knowledge and information between local stakeholders [BAL 08] and promote the diffusion and the adoption of innovation by these stakeholders [LIN 10]. From then on, the state reduced its direct involvement in R&D

and created incentives for the emergence of innovation networks. The measures aimed to encourage greater connectivity between academic research and industry (collaborative transfers mandated by the government) to promote innovative entrepreneurship (incubators, spin-offs) and create a conducive environment for the development of new procurements of innovative products (technical standards, public procurement, etc.).

Thus, the Taiwanese innovation system transformed from a NIS built from scratch by the state to absorb foreign technologies and accumulate local knowledge into a NIS that is based on the interactions and the flows of cognitive and financial information between stakeholders of R&D and engineers to develop the innovative capabilities of domestic firms.

Box 2.1. *From technological catch-up to innovation: the Taiwanese national innovation system*

2.2.3. *Measuring the performance of innovation systems in developing countries*

Since the early 1990s, substantial analysis has been conducted to evaluate, compare and rank innovation systems. These works were justified by two main aspects: one of political interest and the other of search for efficiency [BAL 02]. The creation of conceptual structures around innovation is a central objective of technology policy. Moreover, it seems equally important to acquire knowledge on the efficiency of technological processes. The quality of innovation policy depends on the process of learning from experience and especially by comparison (learning by comparing).

The benchmarking method has grown rapidly in systemic innovation studies so that international comparisons may proceed. Moreover, empirical studies are generally done from a given conceptualization of innovation. Most of the studies done in the South establish technological specialization indices or indicators [ALA 98] by emphasizing technological effort. Others open the way for a broader approach without specifying the variables or key indicators that are essential for their representations [ARO 02].

Two types of studies – descriptive and formalized studies – are often carried out. They are mainly directed toward two objectives. They allow the implementation of purely political strategies and/or aim to enhance the conceptual framework of the approach. If they are of political interest, they are generally done by the European Commission, the OECD, UNCTAD and the World Bank, with the aim of developing indicators and methodology for national policy comparison. These works combine benchmarking techniques and systemic approach to innovation. Benchmarking as a comparison process plays a key role in the promotion of innovation and learning policies [LUN 03]. Learning "by comparing" is used to improve the quality of national policies. According to Lundvall and Tomlinson, this technique is relevant to two main methodological problems. The first is the lack of coordination of technology policies and the second is based on the fragility of international comparisons in terms of the quality of interactions, the presence of learning organizations mainly at the level of tacit factors, which are particularly important in the innovation process.

When the studies do not look into the implementation of policies, they serve more modestly to improve the structure of innovation systems and produce descriptive models. Among these, that of Liu and White [LIU 01] is certainly one of the most striking. These authors identify five activities related to the innovation process: research, production, users (end-users), interaction and education. Their analysis is based on the distribution of activities in the innovation process, on the control and coordination mechanisms and on the information flows. They apply their method to analyze China's innovation system through an inter-temporal comparison of the various stages of development. Differences in institutional, organizational and performance contexts are thus highlighted between autarkic China's innovation system in the 1960s and 1970s and that of today's China, which is more open to competition. Also in this regard, Chang and Chen [CHA 04] compared the innovation systems with other structural forms (technological and sectoral systems) in terms of knowledge. Three drivers are proposed: the production of knowledge, the inter-organizational relationships and the factors facilitating the transfer of knowledge. Several techniques are presented to promote innovation systems in a knowledge economy perspective. These are relative to systemic links and business linkages.

The current Chinese NIS is the result of a series of structural reforms that were initiated in the 1980s. Before the reforms, China's science and technology system was built according to the logic of the Soviet model. It was organized by rigid vertical hierarchical relationships within the Academy of Sciences and ministries as well as by weak horizontal links between R&D stakeholders and production units under the supervision of various ministries. In this rigid system, bureaucratic inconsistencies in the organization of science and technology systems led to the blockage of the stakeholders, that is to say, academics, research institutions, universities and some large companies. The authorities acknowledged that the scientific community lacked creativity because innovation activities were subject to strict and omnipresent state control.

The structural reform of the science and technology system launched in 1985 aimed to correct these failures by building a national innovation system in which Chinese companies could build innovation capabilities by their own initiatives. To do this, it was necessary to intensify the flow of knowledge between stakeholders and thus support the interactions between business, research centers and universities [LUN 12]. The main measures taken were the following:

– Decentralization of science and technology policy that asked local authorities to set their own relative policy while taking into account the guidelines included in the general framework to guide the research and the modernization of the national economy.

– Establishment of a system of contract responsibility and an evaluation system by numerical criteria to improve the effectiveness of public research.

– Reconciliation of public research with industry by encouraging research institutes to provide engineering services to public companies, later extended to private companies.

–Technology commercialization and entrepreneurship by the staff of public research.

The construction of the national innovation system was supported by structured innovation policy to allow Chinese companies to develop their competitiveness by methodically following a well-defined technological trajectory [LIU 14]. The typical Chinese technological development trajectory was characterized by a process of learning and knowledge accumulation to build the innovation capacity of public and private companies. In a first instance, the focus was on the absorption of foreign technology, supported by industrial policies, then imitation, reinvention and finally innovation [LIU 13]. At the heart of this structured approach, the big public companies exerted a leverage effect on directing the industrial development according to the path set by the State. Through planning, the Chinese State mobilized enormous resources that enabled companies in the country to quickly comply (after thirty years) with the Western technological standards.

As the early stages of structural reform (1985–2006) were focused on technological catch-up through the acquisition of foreign technology, new systemic failures emerged. On the one hand, public research took advantage of institutional reforms to develop its own business and move away from industry. On the other hand, supported by foreign investment policies that required, *inter alia*, proactive technology transfer in scientific and technical capital intensive sectors, companies, especially large public companies, remain dependent on foreign technologies.

Thus, in 2006, the Chinese government redefined its national innovation policy for promoting endogenous technological change by Chinese companies. The new measures aimed to better coordinate interministerial and multidisciplinary programs to create a dynamic network between the different stakeholders of innovation: strengthening industry–university–research– training collaboration, mobility of researchers toward industry and technology services, promoting innovative entrepreneurship.

Box 2.2. *Structural reforms and the construction of a national innovation system in China*

The Indian national innovation system is characterized on the one hand by the integration into global knowledge networks and on the other hand, by the development of innovations driven by the need to provide affordable and accessible solutions to poor consumers.

According to the main stream paradigm, innovation capacity is measured by R&D expenditure, number of researchers and number of patents. In this context, India has a high-quality R&D infrastructure. The Indian government initiated the construction of a scientific and technological system based on public–private partnership. Public effort is more concentrated in selected sectors such as nuclear, biotechnology or space. The education system, meanwhile, constitutes a reservoir of highly-skilled English-speaking workforce. The strengthening of the protection of intellectual property in the late 1990s, coupled with the technological advantages that the country offers make India an attractive destination for foreign investments. Specializing in outsourcing of knowledge-intensive activities and the presence of multinational research centers enable India to integrate into global research networks [OEC 12].

A new innovation model has begun to appear in developing countries: frugal innovation. This model is rooted in new combinations of known elements and does not require input of new knowledge: mobility, nutrition, housing, energy, etc. Frugal innovation is widespread in India, where the search for simplicity in manufacturing of products for the poor has led to the emergence of many innovative companies [RAJ 12]. In the frugal innovation process, which is often carried out by small businesses, the collaboration between companies and other innovation stakeholders (research and engineering institutions) is essential for developing new production methods, for designing and launching new products, for reducing costs or for finding new solutions that are more tailored to specific needs. Thus, the new Indian innovation policy focuses on the networking of SMEs in the framework of technological cluster construction [MIT 13].

Box 2.3. *The Indian national innovation system and frugal innovation – a new paradigm*

In a broader scope, studies by Amable, Barré and Boyer [AMA 97] on social innovation systems were based on a descriptive approach of six sub-systems that were inherent in the innovation process: science, technology, industry, human resources, education, training and finance. According to these authors, the interactions between different functions of a business (R&D, production, marketing and finance) and between companies and their sociotechnical and economic environments are part of a mode of regulation and an overall consistency where specific technological relationships with global companies are as important as national R&D and innovation structures.

Beyond descriptive models, formalized models have emerged to more "formally" quantify the innovative performance of national economies. More and more authors seek to build instruments to map the NIS. For example, Furman *et al.* [FUR 02] proposed the concept of "national innovative capacity" as the capacity of a country to produce and commercialize a technology flow under the national infrastructure, the environment of industrial clusters and the nature of links between the two. Each component (infrastructure, environment of clusters and link between the two) is measured by variables constituting a specific regression model. This model is used to build relationships between the endogenous growth theory and systemic approach to innovation through a technique based on broad comparisons of national innovative strengths. Porter and Stern [POR 02] applied this model in 75 countries and ranked them on a technological scale. Arcelus and Nasierowsky [ARC 99] also identified domestic technological capabilities on the basis of a structural equation system based on inputs, outputs and moderators. Their analysis technique resulted in the definition of two groups of countries: technology leaders and other emerging countries in which technological advances were based on importation of innovations from abroad. Following the first empirical tests, the authors recently developed a model with two inputs, two moderators and three outputs. The efficiency of the NIS is based on R&D as well as on the productivity of R&D. The overall idea is to analyze the NIS as an isolated sector of the whole economy.

The way to measure national innovation performance is far from obvious. Carlsson *et al.* [CAR 02] recognized this. As a possible solution to this problem, they particularly advocated a restriction of an analytical level on a product, industry or cluster of industries. At first, they recommended analyzing each major component or sub-system. Then, in a second step, the results of the sub-systems could be combined and evaluated by considering the entire system. The fundamental problem of their analysis was based on the relevance of indicators and technological references. All empirical work on the technological capabilities demonstrated the need to compare national innovative performance. The application of technological development indicators now seems obvious. However, the consistency of these models in developing countries remains to be discussed. The input indicators related to R&D seem inappropriate and inadequate to the problem of some developing countries where R&D is almost non-existent.

2.2.4. *Location strategies of multinational firms and the role of NIS*

Until the early 1980s, innovative activities were generally concentrated in the home countries of the multinational firm [PAT 99]. This is explained by the fact that R&D, considered to be a strategic asset, was managed more securely at national level, especially as the standards and rules on intellectual property were not harmonized then. Initially, the internationalization of R&D was limited to exporting innovative products, the patent assignment and licenses and the production abroad of goods with technological content that became common. These forms of internationalization correspond to the exploitation abroad of specific assets of the firm (asset exploiting), that is to say of innovations produced in the country of origin [ARC 97, DUN 95].

During the 1990s, the globalization of the firms' strategy more markedly concerned R&D activities with a major change which consisted of technological production abroad. Cooperation networks between firms, between firms and/or research centers located in various countries were formed. Open innovation [CHE 03] on networking, working, etc. became a constant phenomenon of globalization. Thus, the presence of firms in the global markets was no longer limited to a commercial and productive implementation but also involved research laboratories that had various vocations: local adaptation of products designed by the parent company, the technology intelligence and/or local innovation, to the design of products for global diffusion. The internationalization of R&D was explained, from the point of view of firms, by the need to adapt the products developed in the home country to the standards of the host country and to the specific characteristics of their demand. But it was also reflected in the willingness of firms to tap into the existing scientific and technical resources in these countries. In contrast, in the context of global competition based on innovation capacity, industrial countries and some developing countries sought to strengthen their NIS to attract foreign investments with high added value. They developed structural attractiveness policies based on scientific and technical resources training that are consistent with the production standards of big companies [UZU 11].

Even though the globalization of R&D was mainly concentrated in industrial countries, it increasingly involved emerging countries [FU 10, LUN 09]. The reasons for the globalization of R&D and the growing importance of emerging countries are numerous, they include: the enhancement of the innovation system of the country associated with the lower cost of scientific and technical human resources; the globalization of intellectual property rights; the increasing demand for technological and sophisticated products in some emerging countries; the international deployment of some companies of these countries, which actively participate in the general technological development in the sectors of microelectronics, telecommunications, pharmaceuticals, aerospace or arms (see examples of China, India, Brazil, etc.).

The process of "continuous innovation" is in fact the main feature of a global firm. It is defined as a company whose organization is integrated by multiple information and financial flows and whose structures are largely decentralized: a network firm with multiple subsidiaries, multiple partnerships of sub- and co-contracting; a firm of large financial amplitude in industry and services with a high scientific and technical potential with significant external effects. It has a strong ability to innovate and constantly change its structure and organization. It takes advantage of the comparative and specific advantages of different places and countries where it operates. It integrates all specific activities performed worldwide into a single "value chain" (R&D, logistics, innovation and financial engineering, manufacture, assembly, marketing, services).

The possibilities for decentralized management that the firm has, added to the attractiveness of countries and/or regions (tax preferences, subsidies, efficient innovation system, abundant human capital, etc.), determine the strategic choice of innovation activities location worldwide (Figure 2.1). As the markets are open, but concentrated, characterized by a differentiated and diversified demand, big companies follow a comprehensive and permanent innovation policy, which leads them to invest in the continuous development of their innovation potential and, at the same time, transfer knowledge and technologies to the environment where they settle. Countries with high scientific and technical potential attract these companies, which, through the strategy of diversification of their portfolio and of their partnership, integrate into the innovation systems of host economies.

The territorial base of the company allows it to be a reservoir of resources (and sometimes a market) to cushion the costs associated with its investments in an economy under constant transformation. But this territorial anchoring depends on the quality of the resources that the country offers in relation to its expectations in terms of innovation and business development. Hence, governments and local authorities need to bring out a rich and competitive innovation system by taking into account the competition-cooperation process that is established internationally between big companies. The goal is to attract foreign investment by considering them as technology transfer and learning vectors to integrate global innovation networks and achieve the "technology leapfrogging".

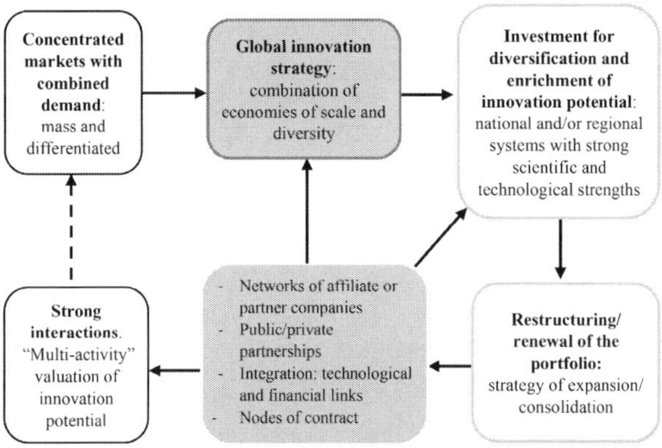

Figure 2.1. *Strategy of innovative activities location of the global firm*

The "technology leapfrogging" in China or India was illustrated by a new stage in the internationalization of innovation strategy of big companies: – the "reverse innovation". To seize opportunities in emerging markets, and be the first on valuable segments in rich countries, companies began to reverse innovation: innovations were designed in emerging countries like China and India and were marketed worldwide [IMM 09]. It was no longer just downstream phases of R&D (product adaptation), but the entire product design process was done in emerging countries. These products were intended for both emerging countries and all countries of the world. Reverse innovation is a strategic response to new global market standards: the development of new markets in emerging countries and the emergence of new competitors from the same countries [LAP 12]. Thus, in many emerging countries there is a large market of potential consumers who are still very poor and thereby ignored by their industrial rivals. Therefore conquering these markets by adapting to their needs and economic constraints could eventually prove to be an interesting challenge. Especially as in some of these countries, a quality workforce has developed, fostering the emergence of local businesses.

2.3. The difficulties of implementing innovation policies in developing countries

While in developed industrial societies, production and use of knowledge take a leading role through the learning process, developing countries remain under the influence of imported knowledge and learning methods without being able to design and implement their own learning processes [ARO 02].

The reasons for the weak performance of the NIS of the South are intrinsic. Besides the lack of adequate human, financial and technological resources, interactions between the innovation potential and the real economy are often non-existent, inadequate or inappropriate. But more than a chronic lack of scientific and technical resources, the strong information asymmetries, due to a failing overall economic governance, prevent the effective promotion of efforts of individual stakeholders of innovation; efforts remain disparate and dispersed.

2.3.1. *Asymmetries and endemic blockages*

Information asymmetries produce poor information and cognitive flows between potential stakeholders of innovation. Universities and R&D laboratories are distant from the production process. Innovation activities are poorly integrated with each other. Isolation of economic stakeholders also appears as a primary reason for the difficulty in building technological capabilities. According to Alcouffe [ALC 92] regarding the Maghreb economies, the lack of integration between research centers and the lack of connections with the production system could be explained by two phenomena: first, the exponential increase in enrolment in universities produced a crowding out of research activities for the benefit of teaching and supervision activities, which involved more abstract training that were increasingly cut off from the productive sector; then, the lack of big industrial groups led to the isolation of production units compared to the research centers and to their suppliers and customers, depriving them of feedback effects on the technological learning process. As for the Latin American economies, this weakness was linked to a lack of producer confidence for local institutions [ARO 02].

The construction of the Algerian research system started in the 1970s, on the eve of independence, based on research conducted by French institutions operating in Algeria. The Ministry of Higher Education and Scientific Research and the National Scientific Research Organization were first created and then completed in the years 1980s and 1990s, by new research institutions [KHE 01, ESA 09]. Efforts were particularly concentrated in the construction of the science and technology base by heavily focusing on human resources: universities and high level research centers. Meanwhile, various structures for the promotion of research and/or networking of innovation stakeholders (engineering, technology centers, etc.) were created for the dissemination of technology to the economy [ASM 14].

The operation of the Algerian NIS however suffered from institutional and structural failures. The promised funds to promote research and development were not granted [KHE 01]. Project selection was subjected to sometimes opaque criteria. Furthermore, as the Algerian NIS is hierarchical, interaction between academia and industry was not ensured. On the one hand, the research system is highly centralized around universities and public research organizations without direct contact with companies [KHE 01, ASM 14]. On the other hand, in the private sector, R&D activities are primarily carried out by a few large companies, while small and medium enterprises (SMEs) do not have access to technology nor capital to finance its modernization. Although Algerian researchers are relatively active in terms of scientific publications (input), the performance of the Algerian NIS measured by the share of high-tech exports, and any patent applications by residents (output) is lower than its neighbors (Box 2.4).

	Intensity of R&D (% of GDP)	Scientists per million inhabitants (FTE)	Technicians per million inhabitants (FTE)	Articles in scientific and technical journals	High-tech exports (% of manufactured exports)	Patent applications by residents
	2005	2005–2012	2005–2012	2011	2012	2012
Algeria	**0.07**	**165**	**33**	**599**	**0.1**	**119**
Egypt	0.24	524	277	2.515	0.6	683
Morocco	0.64	864	53	386	6.4	197
Tunisia	0.92	1.837	43	1.016	5.6	–

(Source: World Bank, *World Development Indicators: Science and Technology*, http://wdi.worldbank.org/table/5.13)

Box 2.4. *The Algerian innovation system: one system, two worlds*

Meanwhile, it is clear that innovation systems in developing countries are incomplete or unbalanced in the sense that certain aspects are inefficient or simply non-existent [NAR 04]. Depending on the nature of the economies studied, we can underline many organizational and institutional rigidities, but also the inadequacy of macroeconomic policies, low investment in education and R&D, difficult integration of innovation activities into the global economy as well as low production capacity and technical assimilation.

> These two countries of Latin America experienced rapid growth in the 1990s, but they were both confronted with the ineffectiveness of their national innovation systems because of institutional or systemic failures.
>
> Although the Mexican government has, since the late 1990s, adopted measures for the development of science and technology, "bad governance" blocks the implementation of innovation policies. Resources devoted to R&D activities are actually less than those declared, and training of research staff remains inadequate. This latter failure is compounded by a lack of interaction between high quality academic research (public and private) and industry. This creates barriers to the dissemination of scientific and technical knowledge and to the development of business innovation capacities [OEC 09].
>
> The Chilean economy is traditionally based on the exploitation of natural resources. Far from major markets and foreign knowledge centers, the Chilean innovation system focuses on public research. Unlike Mexico, the Chilean government has not initiated a broad policy to promote innovation. The majority of R&D activities are funded by the central government and conducted by universities whose relationship with the private sector is poor. The lack of qualified staff and low public support for the dissemination of knowledge and business networking prevent the development of learning and innovation capacity of businesses [OEC 07a].

Box 2.5. *Institutional failures of the innovation systems of Mexico and Chile*

The same applies to sub-Saharan economies: the lack of effort and investment in building technological capacity partly explains the absence of true systemic innovation in these countries. According to Lall [LAL 92], the fragile African economic structure comes from an improper assembly of (1) incentives related to the macroeconomic environment and industrial policies, (2) institutions supporting the national technological structure and (3) capacities related to various resources of technological activity (technical skills, financing and access to local and foreign information). The disconnected nature of these three factors is blocking Southern countries from the industrialization process. These asymmetries are growing and are the result of "bad governance" and, by referring to a set of macroeconomic blockages, reduce the range of decisions and choices for industrial policy and innovation.

But the lack of technological capability can also come from microeconomic elements. According to Ogbu *et al.* [OGB 95], inadequate technological capabilities come from the low production capacity, lack of knowledge transfer vectors between potential stakeholders of innovation, lack of technological learning processes in the organization of the production, as well as a relative lack of inter-company relations. As endemic lack of equipment is one of the characteristics of African underdevelopment, it prevents the building of relationships between companies (and the creation of new businesses); this leads to the inability to achieve, through the effect of technological and managerial learning, incremental or frugal innovations. There are no means for production, no effective links and, therefore, no technological learning or technical change.

This choice of country is based on two key criteria: it is a developing country (also classified by the UN as a "least developed country") and there have been no studies that focused on its systemic innovation. On a macro level, in 2011 Senegal remained very poorly ranked in terms of literacy (170th/182 countries). Nevertheless, its education progresses quantitatively and qualitatively, although the differences between gender and regions still remain. International programs are largely responsible for this improvement (NGOs and associations). In parallel, the knowledge supply has expanded and diversified (technological structures, professional organizations and business support organizations). Higher education has also risen sharply. In 2008, there were 91,359 students of whom 23,318 were in private training institutions. The difficulties of entering the labor market based on the level of education mainly affect higher education graduates. They have the highest unemployment rate. The causes often have to do with recurrent refusing of graduates to accept job offers that are poorly paid and adapted to their training, and with the tendency of non-educated people to accept the first job that is offered to them.

On a microeconomic level, informal micro-enterprise is most representative of Senegal. It has a very small structure (1–2 assets), is not registered with professional institutions and is very often family-owned. The technical and managerial skills are primarily learned through work experience within companies. With external training sources often very poorly available, African companies are forced to almost exclusively rely on training efforts to create their own technical capabilities. In Senegal, a large proportion of business leaders have learnt on-the-job in their own business. The disadvantages of this learning tend to be the lack of knowledge renewal. Indeed, this learning does not increase the efficiency of local businesses. However, it partly contributes to the consolidation and strengthening of individual capacities.

The last aspect to consider: the asymmetry of relations. Not only are relationships between businesses and government complex and ambiguous, but there is widespread opposition between the logic of companies, research institutions and business support institutions. This asymmetry is due in particular to a mismatch between training and economic activities between research and economic priorities, and the neglect of technological concerns in the informal sector. The National Innovation System in Senegal still seems sluggish and unstructured, despite some efforts made in the last twenty years to build national learning capacity.

Box 2.6. *Asymmetries and attempts to create learning capabilities in Senegal*

According to Lall and Wangwe [LAL 98], technological efficiency of sub-Saharan Africa is relatively low and shows little innovation or technological dynamism. Sub-Saharan Africa has lagged far behind technologically and there are only poor signs of modernization. Its structure is traditional, dominated by the production of consumer goods to meet domestic markets. Its capacity to import and use imported technologies for the purpose of technological learning is particularly laborious. While productivity growth is very low, African firms remain below the technical levels of international best technological practices, as well as below levels achieved by other developing countries. Under these conditions, the innovation gap, which is beyond access to technological knowledge, is perceived through a combination of three basic problems [OYE 04]. The first relates to the inability of local institutions to interact with the production stakeholders in order to ensure sufficient autonomous technological dynamics. The second is related to the difficulty in building up local knowledge, through tacit knowledge that is characteristic of small African structures, which failed to meet new challenges arising from the unstable competitive environment. The last stems from repetitive methods of learning by imitation, which creates little opportunities for renewal and modernization of technology used by local businesses. In total, technological exclusion followed by real difficulty in building a learning dynamism limits future economic development prospects.

2.3.2. *The North/South and South/South technology gap*

Polarization or *division of learning* are direct consequences of non-converging trends between the evolutionary trajectory of economies that are built on innovation and knowledge and of economies that rely on technology transfers in order to improve their place in the international division of labor. These polarization effects are directly related to the consequences of globalization and to the increasing speed of technological change. On a large scale, it is clear that the gap between rich and poor countries has been widening continuously. And undoubtedly, the most affected countries remain the least developed countries.

South Africa has undertaken structural reforms since the end of apartheid, opening up to trade and international capital and taking advantage of the improved competitiveness of local products [OEC 07b]. Initially, South Africa inherited a science and technology system with an advanced R&D capacity, which was concentrated on natural resources (minerals and energy), and the military-industrial complexes run by the State. This interventionist and autarkic system was disconnected from the private sector. By maintaining many of its components, reforms since 1994 have aimed to restructure and refocus the science and technology system in order to better integrate it into a national innovation system and push it toward restructuring of industrial bases of the country. The white paper on science and technology in 1996, which served as a guide to reform the science and technology system and construction of the innovation system, in particular raised the issue of the allocation of resources (funding of R&D and innovation, training and mobility of human resources) and the promotion of new governance practices. Thus, institutions were set up to better coordinate the design and implementation of innovation policies [OEC 07b]. After a shift in R&D strategy in 2001, the ten-year program of South Africa of 2008 set a goal of building a knowledge-based economy through the development of human resources and of R&D, by building infrastructure needed for the creation and exploitation of knowledge and the promotion of research development [HAR 13].

With this political will and thanks to clearly-defined innovation strategies, although the national innovation system had to be reinforced, South Africa is seen as the only country among sub-Saharan countries that has a "real" national innovation system with high-quality research institutions where collaboration between universities and industry is encouraged and the private sector contributes to the overall R&D effort [MUG 09]. South Africa accounts for the majority of scientific research measured by the number of patents and scientific publications of all African countries [ONU 07] and it has the most important absorptive capacity of technologies and scientific knowledge among the sub-Saharan countries [MUG 09].

Box 2.7. *The innovation system in South Africa: looking toward the future*

Globalization has contributed to the intensification of competition in all markets, directly impacting on global and national economic dynamics. Production and consumption activities, as well as their components (capital, labor, raw materials, management, information, technology), are organized on an international scale, contributing to diversification and increasing quantities of goods and services that are exchanged. This trend has also led to a transformation in the nature of markets, which are now based on knowledge and learning. The fastest developed production factor in the twentieth century was in fact human capital. This point reflects the importance of intangible factors as key elements of competitiveness. Finally, while internationalization of the economy has facilitated the flow of information and technology at a faster speed and lower cost than in the past, it is also the result of development and use of information and communications technology [GRE 01]. In addition, it has taken part in the rapid development of a society based on information; *know-why* and *know-what*, are based on codified knowledge that is easily transmitted via computer networks.

Another fundamental aspect is the increase in demand for qualifications and skills that are related to the introduction of information technologies. This higher-speed change is itself linked to the rapid development and wide dissemination of new information and communications technology. This acceleration of technological change can be explained by a reduction of the time taken between design and launch of high value-added products; the time taken between production of knowledge and its commercialization is also shortened. All this has resulted in the shortening of product lifecycles. New technologies play a key role in the nature of change. But this holistic and necessarily simplistic view only allows us to present current trends. Also, although it is clear that the development of new knowledge has helped accelerate technological change, it has also widened development inequalities between countries.

Although it is difficult to quantify, ongoing innovation processes generate the need for increased learning capacity so that innovation systems and enterprises in particular can meet new market requirements. Under the constant pressure of competition, companies carefully select their employees according to their skills and carefully choose the country in which they operate. The first decision criterion is the presence of technological capabilities [LUN 97]. In any case, the globalization of markets has not been correlated with uniform diffusion of technology. These are not spread between companies and between countries with the same intensity and with the same amplitude. While industrialized countries have *rich interactive learning spaces* between the transmitters and receivers of information and knowledge, in the South, these spaces consist of weak ties because of their poor potential for innovation [ARO 03].

Innovation and economy of learning assume the existence of capacity to acquire and accumulate both codified and tacit knowledge. It is through the combination of *opportunities* and *learning capacities* that the polarization phenomenon is likely to exacerbate or diminish. In other words, in order to build learning capacity, efforts should be made on the level of demand (new technologies, new skills and competencies, stable macroeconomic environment, proactive innovation policy) and supply (institutional size, organizational and managerial skills, ability to absorb technologies, access to external technical information) [LAL 92]. From this, roughly three broad poles emerge. The first is composed of former and new industrial countries and derives its power from its knowledge-generating scientific and techniques capabilities, from its derived techno-sciences that bring innovation and from a social organization that is formed in symbiosis with scientific and technological development. The second pole consists of a set of countries that have political stability and that implement proactive economic and innovation policies, as it clings to the first pole by acquiring certain competitive advantages (energy, low-tech, financial and tourism services, etc.) that are necessary to hold an intermediate place in globalization. The third pole, devoid of scientific and technical basis, is simply a consumer of scientific knowledge and technological innovations that are developed in the other two poles without really benefiting from them. In the least developed countries, the impact of the *polarization* and *division of learning* is not difficult to see. For example, sub-Saharan Africa sees its structural competitiveness melt even though these countries are well supplied with raw materials (energy, industrial and agricultural). In this case, the fracture is absolute [DOZ 05].

In terms of overall economic data, the following table (Table 2.4) shows the evolution of economic growth rates of the major developing regions of the world. Economic growth in developing countries is sometimes higher than that of Northern economies, particularly in sub-Saharan Africa or South Asia. African growth during the year 2000 was mainly due to investments in infrastructure, increased agricultural production and expansion of the services sector. The worrying situation of European growth is in turn due to the recession of Eastern European economies, which have undergone a long transition period from planned economy to market economy under the conditions of consolidation, and monetary and fiscal austerity.

In terms of technological capabilities in the strictest sense (science and technology), the following figure (Figure 2.2) shows the technological performance of several regions, including industrialized territories (Europe and USA), transition economies (Eastern Europe), East Asia (including China), Latin America and the Caribbean, sub-Saharan Africa (including South Africa), sub-Saharan Africa (excluding South Africa) and North (Maghreb) and East Africa. Four indicators are used to assess the technological performance of the above regions: the industrial added value per capita, the manufacturing exports per capita, the share of medium and high-tech in industrial added value and the same share in exports.

% GDP	1961	1971	1981	1991	2001	2012	2013	2014
East Asia and the Pacific	6.77	5.16	4.55	4.36	2.32	7.4	7.2	6.9
Europe and Central Asia	5.38	3.71	0.46	0.67	2.15	1.9	3.7	2.4
Latin America and the Caribbean	6.35	6.31	1.48	3.71	0.72	2.6	2.5	0.8
Middle East and North Africa		9.84	2.86	3.46	2.12	1.4	0.5	1.2
South Asia	4.11	0.82	6.09	1.86	2.74	5	4.9	5.5
Sub-Saharan Africa	2.67	6.54	1.60	-0.08	3.54	4	4.2	4.5

Table 2.4. *Regional trends of growth in developing countries and countries in transition (source: Compiled from the World Bank, Global Outlook, 2015 for 2012–2014 data and World Development Indicators, 2014 for 1961–2001 data (http: //data.worldbank.org))*

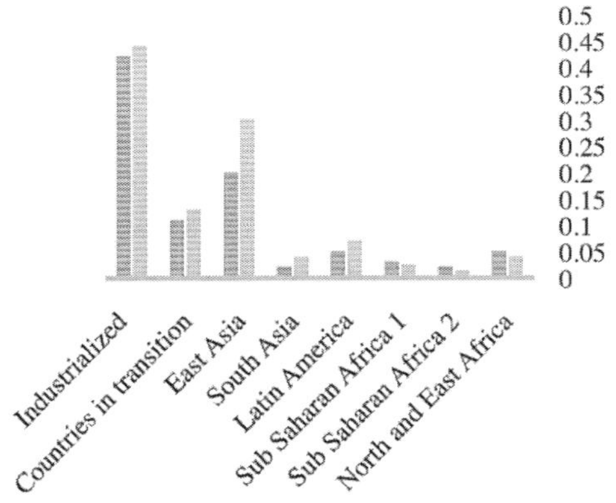

Figure 2.2. *Industrial performance by region (1995/2010) (Source: [UNI 11])*

The industrial performance of sub-Saharan Africa is poor compared to other regions. As Freeman points out [FRE 02, p. 154], "unequal distribution of benefits and social costs occurs on an international scale, with some nations benefiting from new technologies and others unable to do so". According to Lall and Wangwe [LAL 98], the technological potential of sub-Saharan Africa is relatively low and shows little innovation dynamism. Africa has lagged far behind technologically and it only shows poor signs of modernization. Its structure is traditional, dominated by the production of consumer goods to meet domestic markets. Its capacity to import and use imported technologies for the purpose of technological learning is particularly laborious. While productivity growth is very low, African companies remain below technical levels of international technological best practices, as well as below the levels of other developing countries.

Moreover, the role of the technology holders, who are associated with lenders, in international trade, seems to have constrained rather than fostered technological learning. For example, in the last twenty years Latin America has received significant financial assistance to modernize its economy. But its resources were not invested in the training of human capital, leading to the deterioration of structural competitiveness of these countries. The technological gap between Latin America and Asian regions is illustrated by unequal access to resources that are necessary to build technological capabilities. As shown in Table 2.5, sub-Saharan Africa and South and East Asia stand out from other regions at the higher education level. Although global data show a total increase of almost 3 percentage points between 1991 and 2004 of the tertiary school life expectancy, sub-Saharan Africa only grew by 0.7 points in 13 years. The school life expectancy is a synthetic indicator of access to education: it adds tuition at the primary, secondary and tertiary level. It is the probable number of years a child is expected to attend in the school and university system. Globally, and if current enrollment rates remain the same, a child can expect to remain in the education and training system on average 10.7 school years. This goes up to 15.7 years if the child lives in a developed country, 12.7 in a country in transition and goes down to 10.1 years in a developing country, with 7.6 years in sub-Saharan Africa, twice less than in a rich country.

The share of young high school graduates who enroll in college is particularly low in sub-Saharan Africa, while Latin America remains in the world average (Table 2.5).

Region	1991 (%)	2004 (%)
Arab States	6.7	9.6
Central and Eastern Europe	10.2	20.2
Latin America	8	10.3
South and East Asia	3.8	5.6
Sub-Saharan Africa	2.5	3.2
World	7.4	10.2

Table 2.5. *Share of high school graduates pursuing higher education by region (1991 and 2004) (source: UNESCO, Global Education Digest, 2006)*

This is the same for enrollment in higher education worldwide. North America/Australia/New Zealand have the highest share of students in the labor force (9.8% each), followed by Europe (9.1%). Then the Near and Middle East (8.7%), Central and South America (7.7%), which is far ahead of Asia (3.7%) and Africa (7.7%) [OST 10].

The evolution of R&D spending follows the same path (Tables 2.6 and 2.7). In 2007, North America attained 36.2% of global R&D, Asia 31.2% and Europe 27.3%. These three areas account for 95% of global R&D spending and dedicate 1055 billion dollars on research and development. Africa only grew by 0.1 point between 2002 and 2007, representing just 0.7% of global R&D.

Domestic expenditure on R&D	World share 2002 (%)	World share 2007 (%)
Europe	29.8	27.3
Near and Middle East	1.3	1.3
Africa	0.7	0.6
Central and South America	1.8	1.9
Asia	26	31.2
OECD countries	84.7	79

Table 2.6. *Domestic expenditure on R&D (in % share) (Source: OST Report, 2010)*

	1996	1997	2001	2003	2007	2011	2012
OECD countries	2.17	2.19	2.35	2.29	2.29	2.46	2.40
Europe and Central Asia	0.54	0.62	0.59	0.57	0.65	0.71	
High income	2.19	2.22	2.39	2.30	2.28	2.45	2.32
Middle income			0.65	0.74	0.89	1.24	
Upper middle income bracket		0.47	0.67	0.76	0.96	1.35	
Lower middle income bracket					0.62		
Low and middle income			0.65	0.74	0.89	1.24	
BRICS (average)	0.72	0.74	0.93	0.98	1.06	1.24	1.55
South Africa		0.60	0.73	0.79	0.92		
Brazil			1.04	0.96	1.10	1.21	
China	0.57	0.64	0.95	1.13	1.40	1.84	1.98
India	0.63	0.67	0.72	0.71	0.79	0.81	
Russian Federation	0.97	1.04	1.18	1.29	1.12	1.09	1.12

Table 2.7. *Spending on research and development (% of GDP) from 1996 to 2012 (source: World Bank World Development Indicators, http://databank.banquemondiale.org/)*

Excluding industrial countries, we can observe a concentration of R&D effort in BRICS (Brazil, Russia, India, China and South Africa). Indeed, the intensity of research and development activities, measured by the share of R&D expenditure of the GDP is growing rapidly in these countries and is helping to enrich their innovation systems. Thus, in 2012, the number of scientific publications and the number of patents filed in the BRICS exceeded those of all the countries that make up the middle-income countries (Table 2.8).

	Number of scientists (FTE per million inhabitants)	Scientific and technical publications	Use of royalties from intellectual property (million US $)		Patent applications	
			Revenue	Payments	Residents	Non-residents
	2005–2012	2011	2012	2012	2012	2012
World	**1284**	**582,012**	**242,153**	**254,565**	**1 430,327**	**739,805**
High-income countries	3864	407,339	23,646	209,058	853,510	506,349
Middle income Countries	569	173,061	5 557	45,412	568,225	232,881
Upper middle income bracket	897	141,716	4921	37,232	553,837	185,255
Lower middle income bracket	–	31,345	635	818	14,388	47,626
Low and middle income	528	174,673	5693	45,506	576,817	233,456
Low-income countries	–	1612	136	95	–	–
BRICS (total)	**886 (average)**	**142,799**	**2607**	**31,460**	**578,979**	**185,565**
Brazil	710	13,148	511	3666	4804	25,312
China	102	89,894	1044	17,749	535,313	117,464
India	160	22,481	321	399	9 553	34,402
Russia	3096	14,151	664	7629	28,701	1551
South Africa	364	3125	67	2 017	608	6836

Table 2.8. *Science and technology indicators by country group (source: World Bank World Development Indicators, http://databank.banquemondiale.org/)*

The performance of an innovation system is visible through the improvement of the position of these countries in the global innovation ranking (Table 2.9). With the exception of India, which dropped out due to its political instability, the increase was dramatic for two countries: China (67th in 2010 to 29th in 2014, world ranking) and South Africa (99th in 2014 to 53rd in the world in 2010); these countries implement proactive innovation policies to strengthen their national innovation systems (see Boxes 2.2, 2.3 and 2.7).

10 most innovative countries in 2014
Switzerland (1)
United Kingdom (2)
Sweden (3)
Finland (4)
The Netherlands (5)
United States (6)
Singapore (7)
Denmark (8)
Luxembourg (9)
Hong Kong (China) (10)

BRICS countries ranking		
	2014	2010
China	29	67
South Africa	53	99
India	76	56
Brazil	61	68
Russia	49	64

Table 2.9. *The 10 most innovative countries in the world in 2014 and the positioning of emerging countries in the ranking of global innovation (source: The Global Innovation Index, 2009–2010, 2013, 2014)*

The same goes for scientific production. In 2008, 38.6% of scientific publications produced in the world involved at least one European country, 28.4% involved a North American country and 24.3% an Asian country. These three areas account for over 90% of world scientific production. Central America and Africa respectively contribute 2.9% and 1.3%. These results are positively correlated with the number of scientists: more than 2 million in Europe, 2.6 million in Asia versus 36,000 in Africa and 165,000 in Central and South America [OST 10].

Although R&D is crucial in terms of absorption and adaptation of imported technology, many developing countries do not have the specific learning capacities of a modern economy. The example of the African continent demonstrates this easily: its learning capacity is weak and fragile, the "industrialization" (or modernization) of production processes in these countries is impossible. Lall and Pietrobelli [LAL 02] qualify the African technological system as inadequate due to the lack of scientific and technical resources, the absence of a stable and predictable business climate and the difficulty in designing and implementing a basic innovation policy.

2.3.3. *The structural problems of innovation policies*

The successes and failures of developing countries in innovation are partly due to the general conditions in which business activity is conducted: labor market, regulation, tax system, etc. Public resources and capacity management policy is important. The flaws in national economic governance may restrict knowledge diffusion and impede solving specific problems. Moreover, the massive introduction of information and communications technologies in many developing countries contributes to the building of a strong informational system that allows them to capture the various streams of knowledge. This is favorable to countries considered as lagging behind because the international rules of globalization often offer new opportunities for localized learning (via mechanisms of control and application of scientific and technical knowledge).

Innovation policies in developing countries, however, are considered insufficient or inadequate due to numerous reasons. Firstly, rentier economy and "bad governance" prevent the effective appropriation and valorization of local or imported knowledge. The prerequisite to using the knowledge is only targeted technological purposes without favoring local knowledge. The marginalization of indigenous traditional knowledge increases the cost of innovation policies and blocks the emergence of learning opportunities.

To move from a rentier economy founded on primary resource export to a profit-making economy based on the creation of value and long-term competitiveness, one must be concerned with the creation of technological and innovation capabilities. The starting point is improvement of scientific learning and technical capabilities and managerial capacities of existing firms and public or semi-public institutions. To draw this framework for technological resource building and valorization, the political elite must be able to, as a first step, provide leadership on the management of public goods. Such leadership requires a realistic and achievable vision for the development of a country. This then becomes a political issue. Insufficiently qualified policy makers cannot grasp the dynamics of innovation nor identify difficulties and the necessary information for decision making. The State thereby has almost no capacity to design a program for the development of productive capacity of a country. From the moment when the rise of a national economy is not accompanied by a political control on the model, on the structure and the pace of development, all innovation policy formula appears to be unfounded. Creating learning skills presupposes that citizens are nourished, informed and educated. We cannot respond to innovation and knowledge internalization without proper social guarantees. How does one train and manage human resources for development without infrastructure for energy, transport and communication, health, banking, etc. which are sufficiently dense and accessible to the population? And conversely, how does one access basic services with low cash income? Therefore, it is necessary, in order to support the design and development of new training and learning policies, to have new leadership that knows above all that the issue is not just economic but also political and social.

Moreover, political choices are often questionable. Many developing countries opt for the creation of technology parks and the promotion of sectors directly related to science and technology with a view to increasing the attractiveness of the territory and to integrating into global innovation networks. But several major drawbacks limit this new claim. Technological identity cannot be built on non-existent or impaired capacities. Brain drain, lack of planning and resources and ignorance of local potential are among the biggest barriers to innovation. Another point is that knowledge must be used where capacity is and not where the government wants to establish it. "The myth to which we owe this other headlong rush is that while the dissemination of information and communication technologies on a large scale should accelerate development, the education system is so bad that young people trained at great expense in the North remain there or return there" [TRA 02, p. 90]. It would therefore be more appropriate for developing countries to build their own traditional knowledge systems [JUM 01] following the recent example of India focusing on frugal innovation.

Modernization of knowledge thus does not only pass by technological capabilities, it requires above all the capacity and opportunities for promoting endogenous traditional knowledge in a range of sectors (especially in low and medium technologies). Economic emergence in this sense is the emergence of knowledge and daily learning practices in small traditional farms or in small businesses. For example, in many African economies innovation capacity is built in the traditional sectors such as agriculture, mining, crafts, etc. but also in the informal sector. It would therefore be wise to adapt the financing of informal activities to the requirements of the environment, build institutional capacity in the primary products sector [DIA 04] or even help interactive learning in traditional communities before investing in more expensive and risky activities. As for the enhancement of human capital, if this currently is through the construction of research centers or laboratories, a policy of on-the-job training is essential: reinforcement of analytical capabilities and local expertise is necessary to guide the formulation and the implementation of development policies and programs [OLO 00].

Employment schemes created in least developed countries do not favor the relationship between skills, aspirations and employability. "Skills are not always highlighted due to the differences in the technicality and more often, in the social organization of labor" [SMI 03, p. 192]. In terms of training and learning, there are few concrete measures for the integration of the most disadvantaged groups, in order to place them under favorable conditions to the development of their activities. In other words, qualification of the workforce is still largely ensured through informal structures through non-institutionally recognized learning. The answer to the question of who uses human capacity lies in non-governmental organizations (NGOs), which often substitute for the State and contribute to the financing of training activities in small community structures. NGOs seem better able to design and implement diversified training adapted to local needs. These types of organization are also one of the central facilities to take care of the needs of population [OBR 02]. But the problem is not to replace the "on-the-job" learning through institutionalization, but rather to ensure the coexistence of local knowledge with the knowledge related to activities in the competitive sector. This combination of practices (formal and informal, tacit and explicit) can be highly effective in a perspective of endogenization of knowledge where the theoretical, experimental and professional aspects are a major concern [DUC 98]. In the current situation, the State trains the elites, seeks to develop the research sector and tries to create a skilled workforce, but it ignores the demands and expectations of the job market.

Also, many developing countries are lacking more in technical staff than in administrative staff: the workforce is either highly qualified or too unskilled. But the problem especially arises in terms of knowledge management. The scarcity of executives leads to high cost, especially as these are usually expatriates from Northern countries who work in foreign companies or local companies under outsourcing contract or partnership. Tacit knowledge transmitted by *know-why* and *know-how* are neglected. Although the education system is designed to lead to administrative-type jobs, on-the-job training must play a key role in promoting traditional tacit knowledge. This involves the design and funding of State-enterprise training and employment projects. Companies must be encouraged to develop and systematically manage continuous education programs. But this must go through a fair remuneration of local workforce. With equal competencies, foreign executives are better paid than nationals [NIA 02].

Another important factor that limits the implementation of policies is communitarian solidarity. They encourage a withdrawal behavior resulting in lack of initiative, dependency and disempowerment of individuals. Solidarity networks are therefore unfavorable to growth [NIA 02]. The institutional flaws that this communitarian solidarity maintains prove to be serious obstacles to innovation policies. It impacts on the structure of a state and on the economic stakeholders that are supposed to build skills. As discussed by Diallo [DIA 04], solidarity and control of people outweighs their freedom to undertake, their ability to work and their "capability".

Figure 2.3 shows the "vicious circle" of technological dependence of many developing countries (the least industrialized countries in Africa, Central and South Asia and Latin America). These countries, like all others, need technology to produce and more importantly to export in order to limit their international debt. But due to lack of production structures and self-owned innovation capabilities, the flow of imported technology increases with the country's increasing investment needs resulting in growing external debt and weakening competitiveness.

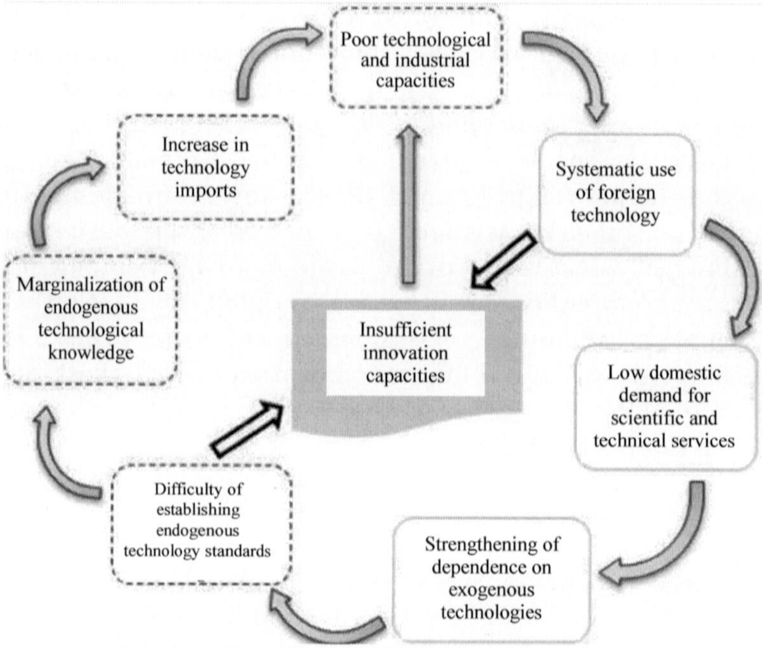

Figure 2.3. *The vicious circle of technological dependency*

The gap between imported technology and the low level of industrialization of the host economy slows internal distribution of productivity gains. As the inter-industrial links are missing, the gap is also widening between the so-called "modern" sector, consumer of imported technologies, and the so-called "traditional" sector consisting of activities of low added value and which is disconnected from the dynamics of capital formation in the country. Such asymmetries increase the difficulties in implementing an overall innovation process due to the limited and outward-oriented modern and competitive sector of an economy.

Developing countries must keep up with the changing criteria of international competition in order to remain in the world economy. The accelerating pace of innovation and the increased role of knowledge in international competition have continued to increase the cost of technological progress. Consequently, many developing countries suffer this "pressure" without intervening to reshape their own economic structures and transform the stress of globalization into an opportunity for development. The State obviously has a central role in the construction of scientific and technical learning capabilities that are necessary for the emergence or reinforcement of the national innovation system.

Conclusion

The challenges of globalization have revived the debate on the possibilities of redefining the role of economic policy in development and via technological, organizational, institutional and social innovation. With the release of the hypothesis of perfect markets and the proliferation of market failures, the study of some theoretical approaches through current changes shows that globalization re-legitimizes an active economic policy that is focused on development and is defined according to a *clever interventionist* approach. Moreover, the analysis of recent practices in political choice proves that it remains realistic, even imperative, in the context of an open and hypercompetitive economy.

Yet less than two decades ago, liberal reforms such as those materialized by the structural adjustment programs and then by the Washington Consensus were still presented as the best treatment to economic, financial and social problems in developing countries. The supremacy of "all market" was then established and less State policy became the rule. Trade liberalization, that of capital markets, privatization, fiscal austerity, elimination of public subsidies, etc., were all reforms intended to allow the "free enterprise" to fully play its alleged role in the promotion of economic and social development. The results of reforms have been mixed. With slower growth, unmitigated macroeconomic imbalances and exacerbated social inequality, the minimum State policy proved to be far from being a panacea for "bad development". Although international institutions had no other choice than to relativize their speeches, emphasizing the need for "good governance", the idea was to complete the first liberal reforms, including institutional reforms, without questioning their relevance or theoretical foundation.

Indeed, in theory, the real breakthrough that fundamentally questioned the appropriateness of the minimum State policy was carried out under the approaches of endogenous growth. By explaining growth through accumulation behavior, these approaches restored the structural economic policy with the solid foundation that had been lacking in the orthodox theory. The endogenization of technological progress in growth models has given the State a central role in encouraging the accumulation of human capital and technological capital, through public policies focused on improving the efficiency of education systems and R&D incentives.

The effectiveness of these policies is now decisive in the development of the structural competitiveness potential of the production system, facing strong dynamic competitive advantages and, more generally, facing the magnitude of challenges of a globalized economy. The paradoxes of globalization, its opportunities and constraints, conflicting goals of its many stakeholders and the differences between public and private or national and global interests are filled with so many mutations that trigger wide innovation processes, according to which individuals, companies, activities and entire economies become marginalized, while others have found opportunities for emergence and development. It is therefore all of these factors that raise the question of the establishment of a global governance system, in order to effectively manage the interactions of globalization that clearly surpass State action and the regulatory power of market. Based on theoretical arguments drawn from the approaches of stakeholders and neo-institutionalists, global governance is an intermediate organizational structure between the State and the market, allowing reconciliation of conflicting interests and objectives, according to a win-win logic between stakeholders of different natures.

However, our analysis shows that the global governance system, as represented by its main stakeholders (WTO, IMF and World Bank), is not the organizational architecture that the theory defines and justifies, based on legitimate decision-making structures and developing equally legitimate purposes, in the view of the overall performance. At least three reasons confirm this finding. The first is that when transposed to the global level, the concept of governance seems withdrawn from the State a growing part of its decision-making powers and presents itself as a sort of *substitute for proactive national economic policies*. The second reason, which derives from the first, is that the objectives of the global governance system are unrealistic. By making liberalization and integration into the world economy an end in itself, the ultimate goal of development converging in the global economy remains utopian. The North/South and South/South gap gets bigger.

The global governance crisis is linked to the problem of the diminishing role of national States. Experience shows that the State must develop proactive economic policies to successfully make globalization "profitable". This requires the redefinition of instruments and modes of action of the State, the invention of new institutions and the implementation of active innovation policies corresponding to the new requirements of development and competitiveness.

Global markets are changing through the competition between companies, but also between production systems in different countries. As these two levels are interactive, business competitiveness depends on the performance of production systems and the power of national production systems depends on the technological capabilities of its industrial and commercial companies. An economy is efficient if its technological base is strong. But as the pace of technological change accelerates, the ability to innovate builds the competitive advantages sought by companies and national production systems.

Economic development based on profound changes in patterns of production and consumption of populations is intimately linked to the capacity of economies to generate innovation. The contribution of innovation systems to the reorganization and to the performance of national economies mobilizes other stakeholders in the economy: State, banks, governments, etc. Innovation policies are, in the current state of competition, the cornerstone of development strategies.

The study of a national innovation system is inextricably linked with the study of the construction of skills through the strengthening of learning capacities of the country in question. But to create learning capabilities, an interactive dimension of human capital formation is essential. This occurs when there are: (1) a variety of links in need of suppliers, producers and service providers; (2) a tendency toward cooperation and collaboration; (3) a large number of stakeholders disseminating information and knowledge that is necessary for the production of goods and services; (4) a propensity willingness to stimulate private initiatives through public demand; (5) a strong local-global formal and/or informal networking. These interactions directly influence business performance and the ability to innovate because they allow a combination of different types of knowledge that in turn will generate new codified or tacit knowledge.

The accumulation of knowledge enables the organization of the innovation process from the "competence building" in a fourfold objective: the development of innovative capacity in industry and services; the development of application, maintenance and improvement engineering of techniques; technological learning based on the absorption of imported knowledge and cooperation with foreign investors; and enhancement of the informal sector as an important source of innovation.

Conceptually, an approach based on innovation systems requires a broad definition of the concept of "innovation". To understand systemic innovation in many developing countries, it is important to think in terms of building a knowledge base rather than the technological result. It is also essential to take into account the specific elements of development trajectories in these countries (scope of informal sector, importance of global macroeconomic environment, importance of social capital, etc.) in order to analyze the opportunities and blockages of the innovation process.

Regarding "how to innovate" and ensure North/South technology transfer, building technological capabilities remains paramount to assimilate and disseminate imported technologies, since multinational firms are seeking innovative environments to increase and enhance their technological capital. The capacity building process passes through the implementation of institutional and organizational efforts but primarily through local learning dynamics for creating new collective knowledge. The informal, the tacit, social links and trust are all features to consider in the learning process in developing countries.

Although each developing country (as is the case for industrialized countries) is characterized by a particular socio-economic structure, it is nevertheless clear that there are a number of recurring elements: institutional weakness, weak interactive links, lack of institutions related to science and technology, etc. This is reflected in their technological and innovation capacities. Innovation plays a central role in today's economic growth and development of all countries. It is increasingly recognized that economic processes are achieved interactively by facilitating relationships between different stakeholders that are sources of innovation (companies, academic research and financing). But innovation is also favored by some types of cultures of stakeholders and by public policies that create a more or less favorable environment for the emergence, production and dissemination of new technologies, new goods and new services. This set of stakeholders and their relationships builds up an innovation system. But all the systems are not identical. They differ in the amount of scientific and technical resources they can mobilize, the density and intensity of relations between their members, by the degree of their performance and by their own institutional shapes. They do not all raise the same opportunities to create and develop the learning of stakeholders. Hence, the importance that at the first step, the innovation policy is initially focused on competence-building and promotion of framework of knowledge application that is compatible with an active policy for economic development and integration into the global economy.

Bibliography

[ALA 98] ALCORTA L., PERES W., "Innovation systems and technological specialization in Latin America and the Caribbean", *Research Policy*, vol. 26, pp. 857–881, 1998.

[ALC 92] ALCOUFFE A., Etudes des systèmes nationaux de recherche et développement: le cas de l'Union Maghreb Arabe, note de synthèse, Université Toulouse I, 1992.

[ALT 09] ALTENBURG T., "Building inclusive innovation systems in developing countries: challenges for IS research", in LUNDVALL B.-A., JOSEPH K.J., CHAMINADE C. *et al.* (eds), *Handbook of Innovation Systems and Developing Countries: Building Domestic Capabilities in a Global Setting*, Edward Elgar, Chentelham, pp. 33–59, 2009.

[AMA 97] AMABLE B., BARRE R., BOYER R., *Les systèmes d'innovation à l'heure de la globalisation*, Economica, Paris, 1997.

[AMI 73] AMIN S., *Le développement inégal, essai sur les formations sociales du capitalisme périphérique*, Editions de Minuit, Paris, 1973.

[AMS 01] AMSDEN A.H., *The Rise of the "Rest": Challenges to the West from Late Industrializing Economies*, Oxford University Press, Oxford, 2001.

[AMS 03] AMSDEN A.H., CHU W.W., *Beyond Late Development: Taiwan's Upgrading Policies*, The MIT Press, Cambridge, 2003.

[ARC 99] ARCELUS F.J., NASIEROWSKY W., "Interrelationships among the elements of national innovation systems: a statistical evaluation", *European Journal of Operational Research*, vol. 119, pp. 235–253, 1999.

[ARC 97] ARCHIBUGI D., MICHIE J. (eds), *Technology, Globalisation and Economic Performance*, Cambridge University Press, Cambridge, 1997.

[ARO 02] AROCENA R., SUTZ J., "Innovation systems and developing countries", *DRUID working paper no. 02-05*, Aalborg School, Denmark, 2002.

[ARO 03] AROCENA R., SUTZ J., *Subdesarollo e innovacion, Navigando contra el viento*, Cambridge University Press, Cambridge, 2003.

[ASM 14] ASMANI A., AMDAOUD M., "La gestion de l'innovation dans les entreprises algériennes: enjeu majeur pour l'obtention d'un avantage concurrentiel durable", *Forum Innovation VI du Réseau de Recherche sur l'Innovation*, Paris-La Défense, 1–3 October 2014.

[ASS 02] ASSIDON E., *Les théories économiques du développement*, La Découverte, Paris, 2002.

[BAL 08] BALAGUER A., LUO Y.L., TSAI M.H. *et al.*, "The rise and growth of a policy-driven economy: Taiwan", in EDQUIST C., HOMMEN L. (eds), *Small Country Innovation Systems. Globalization, Change and Policy in Asia and Europe*, Edward Elgar, Cheltenham, 2008.

[BAL 02] BALZAT M., "The theoretical basis and the empirical treatment of national innovation systems", *Volkswirtschaftliche Diskussionsreihe*, Beitrag, WP Universitat Augburg, Allemagne, vol. 232, pp. 31–70, 2002.

[BEL 97] BELL M., PAVITT K., "Technological accumulation and industrial growth: contrast between developed and developing countries", in ARCHIBUGI D., MICHIE J. (eds), *Technology, Globalization and Economic Performance*, Cambridge University Press, Cambridge, pp. 83–138, 1997.

[BEL 02] BELLO W., *Deglobalization, Ideas for a New World Economy*, Zed Books, London, New York, 2002.

[BER 04] BERR E., COMBARNOUS F., “L’impact du consensus de Washington sur les pays en développement: une évaluation empirique”, *Centre d’économie du développement, document de travail no. 100*, Université Montesquieu-Bordeaux IV, 2004.

[BER 97] BERTHE A., *Pour une stratégie de développement technologique*, Horizon Local, 1997.

[BIE 06] BIENAYME A., *Les grandes questions d’économie contemporaine*, Odile Jacob, Paris, 2006.

[BUC 76] BUCKLEY P., CASSON M., *The Future of Multinational Enterprise*, MacMillan, London, 1976.

[CAR 02] CARLSSON B., JACOBSSON S., HOLMEN M. *et al.*, “Innovation systems: analytical and methodological issues”, *Research Policy*, vol. 31, pp. 233–245, 2002.

[CAS 03] CASSIOLATO J.E., LASTRES H.M., MACIEL M.L., *Systems of Innovation and Development: Evidences from Brazil*, Elgar, London, 2003.

[CHA 04] CHANG Y.-C., CHEN M.-H., “Comparing approaches to systems of innovation: the knowledge perspective”, *Technology and Society*, vol. 26, pp. 17–37, 2004.

[CHE 03] CHESBROUGH H.W., *Open Innovation. The New Imperative for Creating and Profiting from Technology,* Harvard Business School Press, Boston, 2003.

[CLI 03] CLIFT J., “Beyond the Washington consensus”, *Finance and Development*, vol. 40, no. 3, 2003.

[COH 89] COHEN W., LEVINTHAL D., “Innovation and learning: the two faces of R&D”, *Economic Journal*, vol. 99, pp. 569–596, 1989.

[CON 10] CONFEDERATION OF INDIA INDUSTRY, INSEAD, *The Global Innovation Index 2009-2010*, INSEAD, 2010.

[COR 13] CORNELL UNIVERSITY, INSEAD, WIPO, *The Global Innovation Index: the Local Dynamics of Innovation*, Fontainebleau, Ithaca, Geneva, 2013.

[COR 14] CORNELL UNIVERSITY, INSEAD, WIPO, *The Global Innovation Index 2014: the Human Factor in Innovation*, Fontainebleau, Ithaca, Geneva, 2014.

[COU 03] COUSSY J., "Les politiques publiques dans la mondialisation libérale: un retour?", *L'Economie Politique*, no. 17, 2003.

[COZ 09] COZZENS S.E., KAPLINSKI R., "Innovation, poverty and inequality: cause, coincidence or co-evolution?", in LUNDVALL B.-A., JOSEPH K.J., CHAMINADE C. *et al.* (eds), *Handbook of Innovation Systems and Developing Countries: Building Domestic Capabilities in a Global Setting*, Edward Elgar, Cheltenham, pp. 57–82, 2009.

[DES 71] DESTANNE DE BERNIS G., "Les industries industrialisantes et les options algériennes", *Tiers Monde*, vol. 12, no. 47, pp. 545–563, 1971.

[DIA 04] DIALLO M.L., *Le Sénégal, un lion économique? Essai sur la compétitivité d'un pays du Sahel*, Karthala, Paris, 2004.

[DOL 04] DOLLA V.S., "Innovation systems, development and the poor: a comparative analysis of China and India", *Second Globelics Conference*, Pékin, 2004.

[DOZ 05] DOZON J.-P., "L'Afrique dans la fracture nord-sud", in AUBRY M. (ed.), *Agir pour le Sud maintenant! Pour une autre approche des relations nord-sud*, Editions de l'Aube, Paris, 2005.

[DUC 98] DUCATEL K., "Learning and skills in the knowledge economy", *DRUID discussion paper series*, Denmark, 1998.

[DUN 95] DUNNING J.H., NARULA R., "The R&D activities of foreign firms in the United States", *International Studies of Management and Organization*, vol. 25, nos. 1–2, pp. 39–73, 1995.

[EDQ 97] EDQUIST C., *Systems of Innovation, Technologies, Institutions and Organizations*, Pinter, London, 1997.

[EMM 69] EMMANUEL A., *L'Echange inégal*, Maspéro, Paris, 1969.

[ESA 09] ESAU S., KHELFAOUI H., "Mapping research systems in developing countries. Country report: the science and technology system of Algeria", CREST, IRD, France, *UNESCO Forum for Higher Education, Research and Knowledge*, 2009.

[FIT 04] FITOUSSI J.-P., *La démocratie et le marché*, Grasset, Paris, 2004.

[FRE 02] FREEMAN C., "Continental, national and sub-national innovation systems-complementarily and economic growth", *Research Policy*, vol. 31, no. 2, pp. 191–211, 2002.

[FRI 53] FRIEDMAN M., *Essays in Positive Economics*, University of Chicago Press, Chicago, 1953.

[FRI 69] FRIEDMAN M., *Inflation et systèmes monétaires* (1968), Calmann-Lévy, Paris, 1969.

[FU 10] FU X., SOETE L. (eds), *The Rise of Technological Power in the South*, Palgrave Macmillan, Basingstoke, 2010.

[FUR 02] FURMAN J.L., PORTER M., STERN S., "The determinants of national innovative capacity", *Research Policy*, vol. 31, pp. 899–933, 2002.

[FUR 95] FURTADO C., *Le Brésil après le miracle*, Maison des sciences de l'Homme, Paris, 1995.

[GAL 84] GALBRAITH J.K., *La voix des pauvres, ou, ce qu'ils ont à nous dire sur l'économie*, Gallimard, Paris, 1984.

[GRE 01] GREGERSEN B., JOHNSON B., Learning economy, innovation systems and development, Paper prepared for the ESST Converge project, Denmark, 2001.

[HAR 13] HART T., JACOBS P., MHULA A., "Review of South African innovation policy and strategy 1994-2012: innovation for rural development", *RIAT concept paper*, HSRC-EPD, no. 3, 30 July 2013.

[HEN 96] HENDRICKX C., "Problématique du transfert de technologie et nouvelles théories de l'innovation et de la firme", *Revue Région et Développement*, no. 3, 1996.

[HIR 74] HIRSCHMAN A.O., *Stratégies du développement économique*, Editions ouvrières, Paris, 1974.

[HIR 81] HIRSCHMAN A.O., *Essays in Trespassing. Economics to Politics and Beyond*, Cambridge University Press, Cambridge, 1981.

[HOB 95] HOBDAY M., *Innovation in East Asia: the Challenge to Japan*, Edward Elgar, Cheltenham, 1995.

[HOB 04] HOBDAY M., RUSH H., BESSANT J., "Approaching the innovation frontier in Korea: the transition phase to leadership", *Research Policy*, vol. 33, no. 10, pp. 1433–1457, 2004.

[HUM 95] HUMBERT M., "Elargissements théoriques standards des politiques industrielles", *Revue d'Economie Industrielle*, vol. 71, pp. 143–162, 1995.

[IMM 09] IMMELT J.R., GOVINDARAJAN V., TRIMBLE C., "How GE is disrupting itself", *Harvard Business Review*, vol. 87, no. 10, pp. 56–65, 2009.

[JEN 07] JENSEN M.B., JOHNSON B., LORENZ E. *et al.*, "Forms of knowledge and modes of innovation", *Research Policy*, vol. 36, pp. 680–693, 2007.

[JOH 03] JOHNSON B., LUNDVALL B.-A., "National system of innovation and economic development", in MUCHI M., GAMMELTOF T.P., LUNDVALL B.-A. (eds), *Putting Africa First: The Making of African Innovation Systems*, Aalborg University Press, Denmark, 2003.

[JOH 03] JOHNSON J., EDQUIST C., LUNDVALL B., "Economic development and the national innovation system approach", *First Globelics Conference*, Rio de Janeiro, Brazil, 2003.

[JUM 01] JUMA C.F., HONCA D., HUETE-PEREZ J. *et al.*, "Global governance of technology: meeting the needs of developing countries", *International Journal of Technology Management*, vol. 22, nos. 7–8, 2001.

[KAT 82] KATZ R., "The effects of group longevity on project communication and performance", *Administrative Science Quarterly*, vol. 27, 1982.

[KHE 01] KHELFAOUI H., La recherche scientifique en Algérie: initiatives sociales et pesanteurs institutionnelles, Où va l'Algérie?, Institut de recherches et d'études sur le monde arabe et musulman, Aix-en-Provence, 2001.

[KIM 97] KIM L., *Imitation to Innovation. The Dynamics of Korea's Technological Learning*, Harvard University Press, Boston, 1997.

[KOZ 04] KOZUL-WRIGHT R., RAYMENT P., "Globalization reloaded: an UNCTAD pers pective", *United Nations Conference on Trade and Development*, Discussion Paper no. 67, January, 2004.

[KUA 03] KUADA J., "Impact of social ties on innovation and learning in the Africa context", in MUCHIE M., GAMMELTOFT P. (eds), *The Making of Africa Innovation Systems*, Aalborg Press, Aalborg, 2003.

[LAL 92] LALL S., "Technological capabilities and industrialization", *World Development*, vol. 20, no. 2, pp. 165–186, 1992.

[LAL 98] LALL S., WANGWE S., "Industrial policy and industrialisation in Sub-Saharan Africa", *Journal of African Economics*, vol. 7, no. 1, pp. 70–107, 1998.

[LAL 00] LALL S., "Technological change and industrialization in the Asian newly industrializing economies: achievements and challenges", in KIM L., NELSON R. (eds), *Technology, Learning and Innovation*, Cambridge University Press, Cambridge, pp. 13–69, 2000.

[LAL 02] LALL S., PIETROBELLI C., *Falling to Compete. Technology Development and Technology Systems in Africa*, Edward Elgar, Cheltenham, 2002.

[LAP 12] LAPERCHE B., LEFEBVRE G., "The globalization of research and development in industrial corporations: towards 'reverse innovation'? The cases of general electric and renault", *Journal of Innovation Economics*, vol. 10, pp. 53–79, 2012.

[LEW 54] LEWIS A., "Economic development with unlimited supplies of labour", *The Manchester School of Economic and Social Studies*, vol. 22, May 1954.

[LIN 10] LIN G.T.R., SHEN Y.C., CHOU J., "National innovation policy and performance: comparing the small island countries of Taiwan and Ireland", *Technology in Society*, vol. 32, pp. 161–172, 2010.

[LIU 13] LIU H., LIANG D., "A review of clean energy innovation and technology transfer in China", *Renewable and Sustainable Energy Reviews*, vol. 18, pp. 486–498, 2013.

[LIU 01] LIU X., WHITE S., "Comparing innovation systems: a framework and application to China's transitional context", *Research Policy*, vol. 30, no. 6, pp. 1091–1114, 2001.

[LIU 14] LIU Z., "Les politiques de promotion des PME innovantes en Chine", *Marché et Organisations*, no. 21, pp. 57–84, September 2014.

[LUN 88] LUNDVALL B.-A., "Innovation as an interactive process: form user-producer interaction to the national system of innovation", in DOSI G., FREEMAN C., NELSON R. *et al.* (eds), *Technical Change and Economic Theory*, Pinter Publishers, Londres, pp. 349–369, 1988.

[LUN 92]LUNDVALL B.-A., *National Innovation System, Toward a Theory of Innovation and Interactive Learning*, Pinter Publishers, London, 1992.

[LUN 94] LUNDVALL B.-A., JOHNSON B., "The learning economy", *Journal of Industry Studies*, vol. 1, no. 2, pp. 23–43, 1994.

[LUN 97] LUNDVALL B.-A., BORRAS S., The globalizing learning economy: implications for innovation policy, Report based on contributions from seven projects under the TSER program, Commission of the European Union, DG XII, 1997.

[LUN 97] LUNDVALL B.-A., NIELSEN P., "Innovation, learning organizations and industrial relations", *DRUID working paper*, Aalborg School, Denmark, nos. 03–07, 1997.

[LUN 02a] LUNDVALL B.-A., *Innovation, Growth and Social Cohesion, the Danish Model*, Edward Elgar, Cheltelham, 2002.

[LUN 02b] LUNDVALL B.-A., JOHNSON B., ANDERSEN E.S. *et al.*, "National systems of production, innovation and competence building", *Research Policy*, vol. 31, no. 2, pp. 213–231, 2002.

[LUN 03] LUNDVALL B.-A., TOMLINSON M., "International benchmarking as a policy learning tool", in RODRIGUES H.J. (ed.), *The New Knowledge Economy in Europe*, Edward Elgar, Cheltelham, pp. 202–231, 2003.

[LUN 09] LUNDVALL B.-A., JOSEPH J., CHAMINADE C. *et al.*, *Handbook of Innovation Systems and Developing Countries: Building Domestic Capabilities in a Global Setting*, Edward Elgar, Cheltenham, 2009.

[LUN 12] LUNDVALL B.-A., GU S.L., "Le cheminement de la Chine vers l'innovation endogène et la croissance économique", in BIRONNEAU R. (ed.), *China Innovation Inc.*, Les Presses de Science Po, Paris, pp. 213–234, 2012.

[MAR 03] MARTRES J.-L., "De la nécessité d'une théorie des relations internationales, l'illusion paradigmatique", *AFRI, Annuaire Français des Relations Internationales*, vol. 4, 2003.

[MCC 99] MCCORMICK D., "African enterprise clusters and industrialization: theory and reality", *World Development special issue on enterprise clusters*, vol. 27, no. 9, pp. 1531–1551, 1999.

[MIT 13] MITRA S., "Innovation clusters initiative: transforming India's industry clusters for inclusive growth and global competition", *The Global Innovation Index*, Cornell University, Insead, Genève, 2013.

[MUC 03] MUCHIE M., GAMMERLTOFT P., LUNDVALL B.-A., *Putting Africa First: the Making of African Innovation Systems*, Aalborg University Press, Denmark, 2003.

[MUG 09] MUGABE J., Knowledge and innovation for Africa's development: priorities, policies and programs, Prepared for the World Bank Institute, March 2009.

[MYR 57] MYRDAL G., *Economic Theory and Under-Developed Regions*, Duckworth, London, 1957.

[MYR 68] MYRDAL G., *Asian Drama: an Inquiry into the Poverty of Nations*, Penguin Press, London, 1968.

[MYT 03] MYTELKA A., "The dynamics of catching up: the relevance of an innovation system approach in Africa", in MUCHIE M., GAMMELTOFT P., LUNDVALL B.-A. (eds), *Putting Africa First: the Making of African Innovation Systems*, Aalborg University Press, Denmark, pp. 29–43, 2003.

[NAR 02] NARAYAN N., PETESCH P., *Voices of the Poor: from Many Lands*, Oxford University Press, Oxford, 2002.

[NAR 04] NARULA R., "Understanding absorptive capacities in an innovation systems context: consequences for economic and employment growth", *DRUID working paper*, nos. 04-02, Aalborg School, Denmark, 2004.

[NIA 02] NIANG C.I., "Aspects socioculturels de la relance et du développement durable", in DIAGNE A., DAFFE G. (eds), *Le Sénégal en quête d'une croissance durable*, CREA, Karthala, Paris, 2002.

[NOR 90] NORTH D., *Institutions, Institutional Change and Economic Performance*, Cambridge University Press, New York, 1990.

[NUR 53] NURKSE R., *Problems of Capital-Formation in Underdeveloped Countries*, Oxford University Press, Oxford, 1953.

[OBR 02] O'BRIEN D.C., DIOP M., DIOUF M., *La construction de l'état au Sénégal*, Karthala, Paris, 2002.

[OEC 07a] OECD, *OECD Reviews of Innovation Policy: Chili*, OECD Publishing, Paris, 2007.

[OEC 07b] OECD, *OECD Reviews of Innovation Policy: South Africa*, OECD Publishing, Paris, 2007.

[OEC 09] OECD, *OECD Reviews of Innovation Policy: Mexico*, OECD Publishing, Paris, 2009.

[OEC 12] OECD, *STI Outlook*, OECD Publishing, Paris, 2012.

[OGB 95] OGBU O.M., OYEYINKA-OYELARAN B., MLAWA H., "Understanding deindustrialization and technological stagnation in Sub-Saharan Africa: a framework", *Technology Policy and Practice in Africa*, IDRC, Canada, 1995.

[OLO 00] OLOFIN S., "Cadre macroéconomique pour la croissance et l'équité en Afrique", in SALL A. (ed.), *La compétitivité future des économies africaines*, Karthala, 2000.

[ONU 00] ONU, OCDE, BM *et al.*, "Un monde meilleur pour tous: poursuite des objectifs internationaux de développement", *Document déposé lors du Sommet de Copenhague+5*, Genève, 25-30 June 2000.

[OST 10] OST, *Rapport de l'Observatoire des Sciences et Technologies*, Economica, Paris, 2010.

[OYE 04] OYELARAN-OYEYINKA B., Learning and local knowledge institutions in African industry, Discussion paper series, UNU/INTECH, Maastricht, Allemagne, 2004.

[PAT 99] PATEL P., PAVITT K., "Global corporations and national systems of national: who dominates whom?", in ARCHIBUGI D., HOWELLS J., MICHIE J. (eds), *Innovation Policy in a Global Economy*, Cambridge University Press, Cambridge, 1999.

[PER 59] PERROUX F., *Les pôles de développement et le développement intérieur: la coexistence pacifique*, PUG, Grenoble, 1959.

[PET 98] PETITEVILLE F., "Les figures mythiques de l'État dans l'économie en développement", *Revue internationale des sciences sociales*, no. 2587, 1998.

[POR 02] PORTER M.E., STERN S., "National innovative capacity", *The Global Competitiveness Report 2001-2002*, World Economic Forum, Oxford University Press, New York, 2002.

[POS 61] POSNER M., "International trade and technical change", *Oxford Economic Paper*, vol. 13, no. 3, 1961.

[PRE 49] PREBISCH R., *The Economic Development of Latin America and its Principal Problems*, United Nations Publication, New York, 1949.

[RAJ 12] RAJAN Y.S., "Shaping the national innovation system: the Indian perspective", *The Global Innovation Index 2012*, Fontainebleau, Ithaca, Genève, 2012.

[ROD 98] RODRIK D., "Who needs capital-account convertibility?", in FISCHER S. (ed.), *Should the IMF Pursue Capital-Account Convertibility? Essays in International Finance*, Princeton University, vol. 207, 1998.

[ROD 03] RODRIK D., "Growth strategies", *Mimeo*, Harvard University, available at: http://www.ksg.harvard.edu/ rodrik/, 2003.

[ROD 05] RODRIK D., Vers une meilleure perception des institutions, CESifo DICE report, 2005.

[ROG 02] ROGOFF K.S., "Contrôle des flux de capitaux: faut-il garder l'esprit ouvert à cet égard?", *Finances & Développement*, vol. 4, pp. 55–56, 2002.

[ROS 43] ROSENSTEIN-RODAN P.N., "Problems of industrialization of Eastern and South-Eastern Europe", *Economic Journal*, vol. 53, pp. 202–213, 1943.

[ROS 60] ROSTOW W.W., *The Stages of Economic Growth: a Non-Communistic Manifesto*, Cambridge University Press, Cambridge, 1960.

[SAW 87] SAWADOGOA., *Un plan Marshall pour l'Afrique*, L'Harmattan, Paris, 1987.

[SAX 01] SAXENIAN A., HSU Y.-J., "The Silicon Valley-Hsinchu connection: technical communities and industrial upgrading", *Industrial and Corporate Change*, vol. 10, no. 4, pp. 893–920, 2001.

[SCH 99] SCHOSER S., "The institutions defining national systems of innovation: a new taxonomy to analyse the impact of globalization", *EAEP Economy Annual Conference*, Prague, 1999.

[SEG 03] SEGURA O., "Competitiveness, systems of innovation and the learning economy: the forest sector in Costa Rica", *Forest Policy and Economics*, vol. 5, pp. 373–384, 2003.

[SEN 00] SEN A., *Development as Freedom*, Anchor Books, New York, 2000.

[SMI 03] SMITH S., *Nécrologie: pourquoi l'Afrique meurt?*, Calmann-Lévy, Paris, 2003.

[STE 77] STEWARD F., *Technology and Underdevelopment*, MacMillan, London, 1977.

[STI 98] STIGLITZ J., "More instruments and broader goals: moving toward the post-Washington consensus", *WIDER Lecture, Helsinki*, available at: http://www.wider.unu.edu/publications/annual-lectures/annual-lecture-1998.pdf, 1998.

[STI 02] STIGLITZ J., *La grande désillusion*, Fayard, Paris, 2002.

[STI 06] STIGLITZ J., *Un autre monde: contre le fanatisme du marché*, Fayard, Paris, 2006.

[SU 04] SU Y., MU R., CHEN K., *et al.*, Research on National System of Innovation for sustainable development, Second Globelics Conference, Beijing, China, 2004.

[TEE 76] TEECE D.J., *The Multinational Corporation and the Resource Cost of International Technology Transfer*, Ballinger Publishing Company, Cambridge, 1976.

[TRA 02] TRAORE A., *Le viol de l'imaginaire*, Bayard, Paris, 2002.

[TSA 03] TSAY C.-L., LIN J.-P., "Return migration and reversal of brain drain to Taiwan: an analysis of the 1990 census data", in IREDALE R., HAWKSLEY C., CASTLES S. (eds), *Migration in the Asia Pacific*, Edward Elgar, Cheltenham, 2003.

[UN 07] UN, Building Science, Technology and Innovative Systems for Sustainable Development in Africa, United Nations Economic Commission for Africa, ISTD Division, January 2007.

[UNC 12] UNCTAD, Innovation, Technology and South-South Callaboration, Report, United Nations Conference on Trade and Development, 2012.

[UNI 11] UNIDO, Annual Report 2011, United Nations, 2011.

[UZU 05] UZUNIDIS D., "Les pays en développement face au 'Consensus de Washington'. Histoire et avenir", *AFRI, Annuaire français de relations internationales*, vol. 6, 2005.

[UZU 10a] UZUNIDIS D., OTANDO G., "Le renouveau des théories du développement. Institutions et bonne gouvernance", *Recherches internationales*, vol. 90, pp. 9–30, 2010.

[UZU 10b] UZUNIDIS D., PATELIS D., "Le nouveau mercantilisme, la mondialisation et sa crise", *Recherches Internationales*, vol. 88, pp. 19–40, 2010.

[UZU 10c] UZUNIDIS D., YACOUB L., *Gouvernance mondiale et politiques de développement: récusation ou renouveau?*, EUE, Berlin, 2010.

[UZU 10d] UZUNIDIS D., YACOUB L., "La crise de la gouvernance mondiale et la question du développement", *Dynamiques Internationales*, vol. 3, pp. 17–32, 2010.

[UZU 10e] UZUNIDIS D., YACOUB L., "Le renouveau de la politique de développement dans la mondialisation. Aspects et perspectives", *Annuaire Français de Relations Internationales*, vol. 11, 2010.

[UZU 11] UZUNIDIS D., LAPERCHE B., "The new mercantilism and the crisis of the global knowledge economy", *Journal of the Knowledge Economy*, vol. 3, no. 2, pp. 373–392, 2011.

[VER 66] VERNON R, "International investment and international trade in the product cycle", *Quarterly Journal of Economics*, vol. 80, pp. 190–207, 1966.

[WIL 90] WILLIAMSON J., "What Washington means by policy reform?", in WILLIAMSON J. (ed.), *Latin America Adjustment: How Much Has Happened?*, Institute for International Economics, Washington, 1990.

[WIL 94] WILLIAMSON O.E., *Les institutions de l'économie*, Inter Editions, Paris, 1994.

[WOR 87] WORLD BANK, Annual Report, Washington, 1987.

[WOR 09] WORLD BANK, World Development Report: Reshaping Economic Geography, 2009.

Index

Other titles from

iSTE

in

Innovation, Entrepreneurship and Management

2015

MASSOTTE Pierre, CORSI Patrick
Sustainability Calling

CORSI Patrick, NEAU Erwan
Innovation Capability Maturity Model

MAILLARD Pierre
Competitive Quality and Innovation

2014

DUBÉ Jean, LEGROS Diègo
Spatial Econometrics Using Microdata

LESCA Humbert, LESCA Nicolas
Strategic Decisions and Weak Signals

2013

HABART-CORLOSQUET Marine, JANSSEN Jacques, MANCA Raimondo
VaR Methodology for Non-Gaussian Finance

2012

DAL PONT Jean-Pierre
Process Engineering and Industrial Management

MAILLARD Pierre
Competitive Quality Strategies

POMEROL Jean-Charles
Decision-Making and Action

SZYLAR Christian
UCITS Handbook

2011

LESCA Nicolas
Environmental Scanning and Sustainable Development

LESCA Nicolas, LESCA Humbert
Weak Signals for Strategic Intelligence: Anticipation Tool for Managers

MERCIER-LAURENT Eunika
Innovation Ecosystems

2010

SZYLAR Christian
Risk Management under UCITS III/IV

2009

COHEN Corine
Business Intelligence

ZANINETTI Jean-Marc
Sustainable Development in the USA

2008

CORSI Patrick, DULIEU Mike
The Marketing of Technology Intensive Products and Services

DZEVER Sam, JAUSSAUD Jacques, ANDREOSSO Bernadette
Evolving Corporate Structures and Cultures in Asia / Impact of Globalization

2007

AMMI Chantal
Global Consumer Behavior

2006

BOUGHZALA Imed, ERMINE Jean-Louis
Trends in Enterprise Knowledge Management

CORSI Patrick *et al.*
Innovation Engineering: the Power of Intangible Networks